Handbook of Optimization Techniques in Power System Operation and Control

Methods and Application

Handbook of Optimization Techniques in Power System Operation and Control

Methods and Application

Dr. Donchin Parasuraman

Oxford Book Company

Jaipur ▪ New Delhi

Published By

Oxford Book Company

Regd. Off. "Bony Residency", Gate No. 2
Opp. Tilak Public School, Vishweswariya Nagar,
Gopalpura Road, Jaipur - 302018, India
Ph: 0141 - 2761280, Telefax: 2761381
email - oxfordbookjpr@gmail.com
website - oxfordbookcompany.com

Delhi Office: 102, 1st Floor, Satyam House,
4327/3, Ansari Road, Darya Ganj,
New Delhi - 110002, India
Ph: 011-45652440

Distributed by
KEYA INTERNATIONAL PRESS LLP
E-mail: keyainternationalpressllp@gmail.com

First published in India, 2024

ISBN : 9789355240781

Typeset by Shivangi Computers, Jaipur

Printed and bound in India by Replika Press Pvt. Ltd.

Contents

Preface

Determining the maximum power system loadability, as well as preventing the system from being operated close to the stability limits is very important in power systems planning and operation. The application of optimization techniques to power systems security and electricity markets is a rather relevant research area in power engineering. The study of optimization models to determine critical operating conditions of a power system to obtain secure power dispatches in an electricity market has gained particular attention. The Optimal Power Flow (OPF) problem with voltage stability (VS) constraints is a highly nonlinear optimization problem which demands robust and efficient solution techniques. Furthermore, the proper formulation of the VS constraints plays a significant role not only from the practical point of view, but also from the market/system perspective. Thus, a novel and practical OPF-based auction model is proposed that includes a VS constraint based on the singular value decomposition (SVD) of the power flow Jacobian. The newly developed model is tested using realistic systems of up to 1211 buses to demonstrate its practical application. The results show that the proposed model better represents power system security in the OPF and yields better market signals. Furthermore, the corresponding solution technique outperforms previous approaches for the same problem. Other solution techniques for this OPF problem are also investigated. One makes use of a cutting planes (CP) technique to handle the VS constraint using a primal-dual Interior-point Method (IPM) scheme. Another tries to reformulate the OPF and VS constraint as a semidefinite programming (SDP) problem, since SDP has proven to work well for certain power system optimization problems; however, it is demonstrated that this technique cannot be used to solve this particular optimization problem.

This challenge has motivated researchers to come up with Optimal Power Flow (OPF) models that better represent power system security in electricity markets. Particular interest has been given to the incorporation of voltage stability (VS) constraints in the OPF, since this phenomena is believed to be directly associated with many major blackouts experienced around the world during the past decade. Consequently, different OPF models with an emphasis on system security have been proposed, such as Security-constrained OPFs (SC-OPFs) and VS-constrained OPFs (VSC-OPFs). However, further research to improve these models and the corresponding solution techniques is needed, since the large computational burden of these models in the solution of real systems is still a problem. Thus, this book elaborates on the development of an enhanced VSC-OPF model, and a robust and eûcient solution technique that can be used in realistic systems. Determining the maximum power system loadability is very important in order to design preventive actions that help keep the system secure even in the worse contingency scenario (N-1 security criterion). The OPF-based Direct Method (OPFDM) is a very flexible and eûcient optimization technique that has been used to carry out this task. However, the theoretical background that supports the use of this model has not been fully addressed in the literature. Therefore, a full theoretical and numerical analysis, is presented in this thesis to formally prove the equivalency of OPF-DM and Continuation Power Flow (CPF) techniques to determine the maximum power system loadability. The SC-OPF, VSC-OPF, and OPF-DM models have been developed using different optimization techniques, such as multiobjective optimization, successive linear programming, and Interior-point Method (IPM). These techniques have become a powerful tool in power engineering to, for example, minimize costs in an electricity market or to determine/prevent insecure operating conditions of a power system. Semidefinite Optimization (SDP) is a very active research area in mathematical optimization, and it has been applied to hydrothermal coordination and power dispatch problems. However, the particular characteristics of SDP, which could be useful in solving VSC-OPFs have not yet been studied. Therefore, this subject is investigated here to determine whether SDP can be applied to the solution of the VSC-OPF problem.

The Handbook is composed of scientists, engineers, and others who have recognized the importance of Optimization networks in Power System and their enormous potential application in many fields including engineering, medicine and finance. Hope readers will find this Handbook to be of great value to them in their professional lives and research.

—Dr. Donchin Parasuraman

Chapter 1

Electric Power Transmission

Electric-power transmission is the bulk transfer of electrical energy, from generating power plants to electrical substations located near demand centres. This is distinct from the local wiring between high-voltage substations and customers, which is typically referred to as electric power distribution. Transmission lines, when interconnected with each other, become transmission networks. In the US, these are typically referred to as "power grids" or just "the grid", while in the UK the network is known as the "national grid." North America has three major grids: the Western Interconnection; the Eastern Interconnection and the Electric Reliability Council of Texas (or ERCOT) grid.

Historically, transmission and distribution lines were owned by the same company, but starting in the 1990s many countries have liberalised the regulation of the electricity market in ways that have led to the separation of the electricity transmission business from the distribution business.

Transmission lines mostly use high-voltage three-phase alternating current (AC), although single phase AC is sometimes used in railway electrification systems. High-voltage direct-current (HVDC) technology is used for greater efficiency in very long distances (typically hundreds of miles (kilometres), or in submarine power cables (typically longer than 30 miles (50 km). HVDC links are also used to stabilise against control problems in large power distribution networks where sudden new loads or blackouts in one part of a network can otherwise result in synchronisation problems and cascading failures.

Electricity is transmitted at high voltages (110 kV or above) to reduce the energy lost in long-distance transmission. Power is usually transmitted through overhead power lines. Underground power transmission has a significantly higher cost and greater operational limitations but is sometimes used in urban areas or sensitive locations.

Figure : 400 kV high-tension transmission lines near Madrid

A key limitation in the distribution of electricity is that, with minor exceptions, electrical energy cannot be stored, and therefore must be generated as needed. A sophisticated system of control is therefore required to ensure electric generation very closely matches the demand. If supply and demand are not in balance, generation plants and transmission equipment can shut down which, in the worst cases, can lead to a major regional blackout, such as occurred in the US Northeast blackouts of 1965, 1977, 2003, and in the west, one in 1996 and 2011. To reduce the risk of such failures, electric transmission networks are interconnected into regional, national or continental wide networks thereby providing multiple redundant alternate routes for power to flow should (weather or equipment) failures occur. Much analysis is done by transmission companies to determine the maximum reliable capacity of each line (ordinarily less than its physical or thermal limit) to ensure spare capacity is available should there be any such failure in another part of the network.

Overhead Power Line

An overhead power line is an electric power transmission line suspended by towers or utility poles. Since most of the insulation is provided by air, overhead power lines are generally the lowest-cost method of transmission for large quantities of electric energy. Towers for support of the lines are made of wood (as-grown or laminated), steel (either lattice structures or tubular poles), concrete, aluminium, and occasionally reinforced plastics. The bare wire conductors on the line are generally made of aluminium (either plain or reinforced with steel, or sometimes composite materials), though some copper wires are used in medium-voltage distribution and low-voltage connections to customer premises. A major goal of overhead power line design is to maintain adequate clearance between energised conductors and the ground so as to prevent dangerous contact with the line. Today overhead lines are routinely operated at voltages exceeding 765,000 volts between conductors, with even higher voltages possible in some cases.

Classification by Operating Voltage

Overhead power transmission lines are classified in the electrical power industry by the range of voltages:

- Low voltage – less than 1000 volts, used for connection between a residential or small commercial customer and the utility.
- Medium Voltage (Distribution) – between 1000 volts (1 kV) and to about 33 kV, used for distribution in urban and rural areas.
- High Voltage (subtransmission less than 100 kV; subtransmission or transmission at voltage such as 115 kV and 138 kV), used for sub-transmission and transmission of bulk quantities of electric power and connection to very large consumers.
- Extra High Voltage (transmission) – over 230 kV, up to about 800 kV, used for long distance, very high power transmission.
- Ultra High Voltage – higher than 800 kV.

Structures

Structures for overhead lines take a variety of shapes depending on the type of line. Structures may be as simple as wood poles directly set in the earth, carrying one or more cross-arm beams to support conductors, or "armless" construction with conductors supported on insulators attached to the side of the pole. Tubular steel poles are typically used in urban areas. High-voltage lines are often carried on lattice-type steel towers or pylons. For remote areas, aluminium towers may be placed by helicopters. Concrete poles have also been used. Poles made of reinforced plastics are also available, but their high cost restricts application.

Each structure must be designed for the loads imposed on it by the conductors. The weight of the conductor must be supported, as well as dynamic loads due to wind and ice accumulation, and effects of vibration. Where conductors are in a straight line, towers need only resist the weight since the tension in the conductors approximately balances with no resultant force on the structure. Flexible conductors supported at their ends approximate the form of a catenary, and much of the analysis for construction of transmission lines relies on the properties of this form.

A large transmission line project may have several types of towers, with "tangent" ("suspension" or "line" towers, UK) towers intended for most positions and more heavily constructed towers used for turning the line through an angle, dead-ending (terminating) a line, or for important river or road crossings. Depending on the design criteria for a particular line, semi-flexible type structures may rely on the weight of the conductors to be balanced on both sides of each tower. More rigid structures may be intended to remain standing even if one or more conductors is broken. Such structures may be installed at intervals in power lines to limit the scale of cascading tower failures.

Foundations for tower structures may be large and costly, particularly if the ground conditions are poor, such as in wetlands. Each structure may be stabilised considerably by the use of guy wires to counteract some of the forces applied by the conductors.

Power lines and supporting structures can be a form of visual pollution. In some cases the lines are buried to avoid this, but this "undergrounding" is more expensive and therefore not common.

For a single wood utility pole structure, a pole is placed in the ground, then three crossarms extend from this, either staggered or all to one side. The insulators are attached to the crossarms. For an "H"-type wood pole structure, two poles are placed in the ground, then a crossbar is placed on top of these, extending to both sides. The insulators are attached at the ends and in the middle. Lattice tower structures have two common forms. One has a pyramidal base, then a vertical section, where three crossarms extend out, typically staggered. The strain insulators are attached to the crossarms. Another has a pyramidal base, which extends to four support points. On top of this a horizontal truss-like structure is placed.

Insulators

Insulators must support the conductors and withstand both the normal operating voltage and surges due to switching and lightning. Insulators are broadly classified as either pin-type, which support the conductor above the structure, or suspension type, where the conductor hangs below the structure. The invention of the strain insulator was a critical factor in allowing higher voltages to be used. At the end of the 19th century, the limited electrical strength of telegraph-style pin insulators limited the voltage to no more than 69,000 volts. Up to about 33 kV (69 kV in North America) both types are commonly used. At higher voltages only suspension-type insulators are common for overhead conductors. Insulators are usually made of wet-process porcelain or toughened glass, with increasing use of glass-reinforced polymer insulators. However, with rising voltage levels and changing climatic conditions, polymer insulators (silicone rubber based) are seeing increasing usage. China has already developed polymer insulators having a highest system voltage of 1100kV and India is currently developing a 1200kV (highest system voltage) line which will initially be charged with 400kV to be upgraded to a 1200kV line.

Figure : Power lines with ceramic insulators in California

Suspension insulators are made of multiple units, with the number of unit insulator disks increasing at higher voltages. The number of disks is chosen based on line voltage, lightning withstand requirement, altitude, and environmental factors such as fog, pollution, or salt spray. In cases where these conditions are suboptimal, longer insulators must be used. Longer insulators, with longer creepage distance for leakage current, are required in these cases. Strain insulators must be strong enough mechanically to support the full weight of the span of conductor, as well as loads due to ice accumulation, and wind.

Porcelain insulators may have a semi-conductive glaze finish, so that a small current (a few milliamperes) passes through the insulator. This warms the surface slightly and reduces the effect of fog and dirt accumulation. The semiconducting glaze also ensures a more even distribution of voltage along the length of the chain of insulator units.

Polymer insulators by nature have hydrophobic characteristics providing for improved wet performance. Also, studies have shown that the specific creepage distance required in polymer insulators is much lower than that required in porcelain or glass. Additionally, the mass of polymer insulators (especially in higher voltages) is approximately 50% to 30% less than that of a comparative porcelain or glass string. Better pollution and wet performance is leading to the increased use of such insulators.

Insulators for very high voltages, exceeding 200 kV, may have grading rings installed at their terminals. This improves the electric field distribution around the insulator and makes it more resistant to flash-over during voltage surges.

Conductors

Aluminium conductors reinforced with steel (known as ACSR) are primarily used for medium and high voltage lines and may also be used for overhead services to individual customers. Aluminium cable is used because it has about half the weight of a comparable resistance copper cable (though larger diameter due to lower fundamental conductivity), as well as being cheaper. Some copper cable is still used, especially at lower voltages and for grounding.

While larger conductors may lose less energy due to lower electrical resistance, they are more costly than smaller conductors. An optimisation rule called *Kelvin's Law* states that the optimum size of conductor for a line is found when the cost of the energy wasted in the conductor is equal to the annual interest paid on that portion of the line construction cost due to the size of the conductors. The optimisation problem is made more complex due to additional factors such as varying annual load, varying cost of installation, and by the fact that only definite discrete sizes of cable are commonly made.

Since a conductor is a flexible object with uniform weight per unit length, the geometric shape of a conductor strung on towers approximates that of a catenary. The sag of the conductor (vertical distance between the highest and lowest point of the curve) varies depending on the temperature.

A minimum overhead clearance must be maintained for safety. Since the temperature of the conductor increases with increasing heat produced by the current through it, it is sometimes possible to increase the power handling capacity (uprate) by changing the conductors for a type with a lower coefficient of thermal expansion or a higher allowable operating temperature.

Power lines sometimes have spherical markers "of one colour" to meet International Civil Aviation Organisation recommendations.

Bundle Conductors

Bundle conductors are used to reduce corona losses and audible noise. Bundle conductors consist of several conductor cables connected by non-conducting spacers. For 220 kV lines, two-conductor bundles are usually used, for 380 kV lines usually three or even four. American Electric Power is building 765 kV lines using six conductors per phase in a bundle. Spacers must resist the forces due to wind, and magnetic forces during a short-circuit.

Bundle conductors are used to increase the amount of current that may be carried in a line. Due to the skin effect, ampacity of conductors is not proportional to cross section, for the larger sizes. Therefore, bundle conductors may carry more current for a given weight.

A bundle conductor results in lower reactance, compared to a single conductor. It reduces corona discharge loss at extra high voltage (EHV) and interference with communication systems. It also reduces voltage gradient in that range of voltage.

As a disadvantage, the bundle conductors have higher wind loading.

Circuits

Figure : Single 3-phase circuit carried by electricity pylon, with ground wire

A *single-circuit transmission line* carries conductors for only one circuit. For a three-phase system, this implies that each tower supports three conductors.

A *double-circuit transmission line* has two circuits. For three-phase systems, each tower supports and insulates six conductors. Single phase AC-power lines as used for traction current have four conductors for two circuits. Usually both circuits operate at the same voltage. In HVDC systems typically two conductors are carried per line, but rarely only one pole of the system is carried on a set of towers.

In some countries like Germany most power lines with voltages above 100 kV are implemented as double, quadruple or in rare cases even hexuple power line as rights of way are rare. Sometimes all conductors are installed with the erection of the pylons; often some circuits are installed later. A disadvantage of double circuit transmission lines is that maintenance works can be more difficult, as either work in close proximity of high voltage or switch-off of 2 circuits is required. In case of failure, both systems can be affected.

The largest double-circuit transmission line is the Kita-Iwaki Powerline.

Ground Wires

Overhead power lines are often equipped with a ground conductor (shield wire or overhead earth wire). A ground conductor is a conductor that is usually grounded (earthed) at the top of the supporting structure to minimise the likelihood of direct lightning strikes to the phase conductors. The ground wire is also a parallel path with the earth for fault currents in earthed neutral circuits. Very high-voltage transmission lines may have two ground conductors. These are either at the outermost ends of the highest cross beam, at two V-shaped mast points, or at a separate cross arm. Older lines may use surge arrestors every few spans in place of a shield wire; this configuration is typically found in the more rural areas of the United States. By protecting the line from lightning, the design of apparatus in substations is simplified due to lower stress on insulation. Shield wires on transmission lines may include optical fibres (Optical ground wire (OPGW), used for communication and control of the power system.

At some HVDC powerlines, the ground wire is used as electrode line, connecting the inverter with the ground electrode in order to allow the usage of the ground as conductor for one pole. In this case the ground conductor is mounted on small insulators bridged by lightning arrestors above the conductors. The insulation is required in order to prevent electrochemical corrosion of the pylon. Medium-voltage distribution lines may have the grounded conductor strung below the phase conductors to provide some measure of protection against tall vehicles or equipment touching the energised line, as well as to provide a neutral line in Wye wired systems. Nevertheless also some medium-voltage distribution lines use like high-voltage transmission lines one or two shield wires.

On some powerlines for very high voltages in former Soviet Union, the ground wire is used for PLC-radio systems and mounted on insulators at the pylons.

Insulated Conductors

While overhead lines are usually bare conductors, overhead insulated cables are rarely used, usually for short distances (less than a kilometre). Insulated cables can be directly fastened to structures without insulating supports. An overhead line with bare conductors insulated by air is typically less costly than a cable with insulated conductors.

A more common approach is "covered" line wire. It is treated as bare cable, but often is safer for wildlife, as the insulation on the cables increases the likelihood of a large wing-span raptor to survive a brush with the lines, and reduces the overall danger of the lines slightly. These types of lines are often seen in the eastern United States and in heavily wooded areas, where tree-line contact is likely. The only pitfall is cost, as insulated wire is often costlier than its bare counterpart. Many utility companies implement covered line wire as jumper material where the wires are often closer to each other on the pole, such as an underground riser/Pothead, and on reclosers, cutouts and the like.

Compact Transmission Lines

A compact overhead transmission line requires a smaller right of way than a standard overhead powerline. Conductors must not get too close to each other. This can be achieved either by short span lengths and insulating crossbars, or by separating the conductors in the span with insulators. The first type is easier to build as it does not require insulators in the span, which may be difficult to install and to maintain.

Compact transmission lines may be designed for voltage upgrade of existing lines to increase the power that can be transmitted on an existing right of way.

Low Voltage

Low voltage overhead lines may use either bare conductors carried on glass or ceramic insulators or an aerial bundled cable system. The number of conductors may be anywhere between four (three phase plus a combined earth/neutral conductor-a TN-C earthing system) up to as many as six (three phase conductors, separate neutral and earth plus street lighting supplied by a common switch).

Train Power

Overhead lines or overhead wires are used to transmit electrical energy to trams, trolleybuses or trains. Overhead line is designed on the principle of one or more overhead wires situated over rail tracks. Feeder stations at regular intervals along the overhead line supply power from the high voltage grid. For some cases low-frequency AC is used, and distributed by a special traction current network.

Further Applications

Overhead lines are also occasionally used to supply transmitting antennas, especially for efficient transmission of long, medium and short waves. For this purpose a staggered array line is often used. Along a staggered array line the conductor cables for the supply of the earth net of the transmitting antenna are attached on the exterior of a ring, while the conductor inside the ring, is fastened to insulators leading to the high voltage standing feeder of the antenna.

Usage of Area under Overhead Power Lines

Use of the area below an overhead line is restricted because objects must not come too close to the energised conductors. Overhead lines and structures may shed ice, creating a hazard. Radio reception can be impaired under a power line, due both to shielding of a receiver antenna by the overhead conductors, and by partial discharge at insulators and sharp points of the conductors which creates radio noise. In the area surrounding overhead lines it is dangerous to risk interference; e.g. flying kites or balloons, using ladders or operating machinery. Overhead distribution and transmission lines near airfields are often marked on maps, and the lines themselves marked with conspicuous plastic reflectors, to warn pilots of the presence of conductors.

Construction of overhead power lines, especially in wilderness areas, may have significant environmental effects. Environmental studies for such projects may consider the effect of brush clearing, changed migration routes for migratory animals, possible access by predators and humans along transmission corridors, disturbances of fish habitat at stream crossings, and other effects.

History

The first transmission of electrical impulses over an extended distance was demonstrated on July 14, 1729 by the physicist Stephen Gray, in order to show that one can transfer electricity by that method. The demonstration used damp hemp cords suspended by silk threads (the low resistance of metallic conductors not being appreciated at the time). However the first practical use of overhead lines was in the context of telegraphy. By 1837 experimental commercial telegraph systems ran as far as 13 miles (20 km). Electric power transmission was accomplished in 1882 with the first high voltage transmission between Munich and Miesbach. 1891 saw the construction of the first three-phase alternating current overhead line on the occasion of the International Electricity Exhibition in Frankfurt, between Lauffen and Frankfurt. In 1912 the first 110 kV-overhead power line entered service followed by the first 220 kV-overhead power line in 1923. In the 1920s RWE AG built the first overhead line for this voltage and in 1926 built a Rhine crossing with the pylons of Voerde, two masts 138 meters high.

In Germany in 1957 the first 380 kV overhead power line was commissioned (between the transformer station and Rommerskirchen). In the same year the overhead line traversing of the Strait of Messina went into service in Italy, whose pylons served the Elbe crossing 1. This was used as the model for the building of the Elbe crossing 2 in the second half of the 1970s which saw the construction of the highest overhead line pylons of the world. Starting from 1967 in Russia, and also in the USA and Canada, overhead lines for voltage of 765 kV were built. In 1982 overhead power lines were built in Russia between Elektrostal and the power station at Ekibastusz, this was a three-phase alternating current line at 1150 kV (Powerline Ekibastuz-Kokshetau). In 1999, in Japan the first powerline designed for 1000 kV with 2 circuits were built, the Kita-Iwaki Powerline. In 2003 the building of the highest overhead line commenced in China, the Yangtze River Crossing.

Undergrounding

Undergrounding refers to the replacement of overhead cables providing electrical power or telecommunications, with underground cables. This is typically performed for aesthetic purposes but increases the costs of electric power transmission and distribution.

Comparison

The aerial cables that carry high-voltage electricity and are supported by large pylons are generally considered an unattractive feature of the countryside. Underground cables can transmissit power across densely populated or areas where land is costly or environmentally or aesthetically sensitive. Underground and underwater crossings may be a practical alternative for crossing rivers.

Other advantages include:

- Less subject to damage from severe weather conditions (mainly lightning, wind and freezing)
- Greatly reduced emission, into the surrounding area, of electromagnetic fields (EMF). EMF consist of both electric and magnetic fields, Due to the electrical shielding (also known as screening) of underground transmission cables they do not exhibit an external electric field. The electric current in the cable conductor produces a magnetic field, but the closer grouping of power cables reduces the resultant external magnetic field and further magnetic shielding may be provided at additional cost.
- Underground cables need a narrower surrounding strip of about 1–10 meters to install (up to 30m for 400kV cables during construction), whereas an overhead line requires a surrounding strip of about 20–200 meters wide to be kept permanently clear for safety, maintenance and repair.
- Underground cables pose no hazard to low flying aircraft or to wildlife, and are significantly safer as they pose no shock hazard (except to the unwary digger).
- Much less subject to conductor theft, illegal connections, sabotage, and damage from armed conflict.

Disadvantages include:

- Undergrounding is more expensive, since the cost of burying cables at transmission voltages is several times greater than overhead power lines, and the life-cycle cost of an underground power cable is two to four times the cost of an overhead power line. Above ground lines cost around $10 per foot and underground lines cost in the range of $20 to $40 per foot. In highly urbanized areas the cost of underground transmission can be 10-14 times as expensive as overhead.

- Whereas finding and repairing overhead wire breaks can be accomplished in hours, underground repairs can take days or weeks, and for this reason redundant lines are run.
- Underground power cables, due to their proximity to earth, cannot be maintained live, whereas overhead power cables can be.
- Operations are more difficult since the high reactive power of underground cables produces large charging currents and so makes voltage control more difficult.

The advantages can in some cases outweigh the disadvantages of the higher investment cost, and more expensive maintenance and management.

Japan

Most electrical power in Japan is still provided by aerial cables. In Tokyo's 23 wards, according to Japan's Construction and Transport Ministry, just 7.3 percent of cables were laid underground as of March 2005.

Europe

The UK regulator Office of Gas and Electricity Markets (OFGEM) permits transmission companies to recoup the cost of some undergrounding in their prices to consumers. The undergrounding must be in National Parks or designated Areas of Outstanding Natural Beauty (AONB) to qualify. The most visually intrusive overhead cables of the core transmission network are excluded from the scheme. Some undergrounding projects are funded by the proceeds of the national lottery.

All low and medium voltage electrical power (<50 kV) in the Netherlands is now supplied underground. Other EU countries such as the UK and Germany are undergrounding a proportion of these cables each year.

California

In the United States, the California Public Utilities Commission's (CPUC) Rule 20 permits the undergrounding of electrical power cables under certain situations. Rule 20A projects are paid for by all customers of the utility companies. Rule 20B projects are partially funded this way and cover the cost of an equivalent overhead system. Rule 20C projects enable property owners to fund the undergrounding.

Variants

A compromise between undergrounding and using overhead lines is installing air cables. Aerial cables are insulated cables spun between poles and used for power transmission or telecommunication services. An advantage of aerial cables is that their insulation removes the danger of electric shock (unless the cables are damaged). Another advantage is that they forgo the costs-particularly high in rocky areas-of burying. The disadvantages of aerial cables are that they have the same aesthetic issues as standard overhead lines and that they can be affected by storms. However if the insulation is not destroyed during pylon failure or when hit by a tree, there is no interruption of service. Electrical hazards are minimised and re-hanging the cables may be possible without power interruption. When undergrounding a powergrid for aesthetic reasons, it is a good idea also to replace open-air switchyards by indoor switchyards and especially pylon transformers by indoor transformers as transformers on pylons, which are only fed by underground cable look somewhat weird.

History of Electric Power Transmission

The history of the technology of moving electricity far from where it was generated dates from the late 19th century. This includes movement of electricity in bulk (formally referred to as "transmission), and the delivery of electricity ("distribution") to individual customers. The distinction between the two terms did not exist in early years and were used interchangeably.

Early Transmission : Prior to electricity, various systems have been used for transmission of power across large distances. Chief among them were telodynamic (cable in motion), pneumatic (pressurised air), and hydraulic (pressurised fluid) transmission. Cable cars were the most frequent example of telodynamic transmission, whose lines could extend for several miles for a single section. Pneumatic transmission was used for city power transmission systems in Paris, Birmingham, Rixdorf, Offenbach, Dresden and Buenos Aires at the beginning of the twentieth century. Cities in the 19th century also used hydraulic transmission using high pressure water mains to deliver power to factory motors. London's system delivered 7000 hp (5 Megawatts) over a 180-mile (290 km) network of pipes carrying water at 800psi. These systems were replaced by cheaper and more versatile electrical systems, but by the end of the 19th century, city planners and financiers well aware of the benefits, economics, and process of establishing power transmission systems.

In the early days electric power usage, widespread transmission of electric power had two obstacles. Firstly, devices requiring different voltages required specialised generators with their own separate lines. Street lights, electric motors in factories, power for streetcars and lights in homes are examples of the diversity of devices with voltages requiring separate systems. Secondly, generators had to be relatively nearby their loads (a mile or less for low voltage devices). It was known that long distance transmission was possible the higher the voltage was raised, so both problems could be solved if transforming voltages could be cheaply performed from a single universal power line.

Specialised Systems

Figure : Streetcars created enormous demand for early electricity. This Siemens Tram from 1884 required 500 volt direct current, which was typical.

Much of early electricity was direct current, which could not easily be increased or decreased in voltage either for long-distance transmission or for sharing a common line to be used with multiple types of electric devices. Companies simply ran different lines for the different classes of loads their inventions required, for example, Charles Brush's New York arc light systems required up to 10,000 volts for many lamps in a series circuit, Edison's incandescent lights used 110 volts, streetcars built by Siemens or Sprague required large motors in the 500 volt range, whereas industrial motors in factories used still other voltages. Due to this specialisation of lines and because transmission was so inefficient that generators needed to be close by their loads, it seemed at the time that the industry would develop into what is now known as a distributed generation system with large numbers of small generators located nearby their loads.

Early High Voltage and Commercial Systems

High voltage was of interest to early researchers working on the problem of transmission over distance. They know from elementary electricity principle that the same amount of power could be transferred on a cable by doubling the voltage and halving the current. Due to Joule's Law, they also knew that the capacity of a wire is proportionate to the square of the current travelling on it, regardless the voltage, and so by doubling the voltage, the same cable would be capable of transmitting the same amount of power four times the distance.

At the Paris Exposition of 1878, electric arc lighting had been installed along the Avenue de l'Opera and the Place de l'Opera, using electric Yablochkov arc lamps, powered by Zénobe Gramme alternating current dynamos. Yablochkov candles required high voltage, and it was not long before experimenters reported that the arc lights could be powered on a 7-mile (11 km) circuit. Within a decade scores of cities would have lighting systems using a central power plant that provided electricity to multiple customers via electrical transmission lines. These systems were in direct competition with the dominant gaslight utilities of the period.

The idea of investing in a central plant and a network to deliver energy produced to customers who pay a recurring fee for service was familiar business model for investors: it was identical to the lucrative gaslight business, or the hydraulic and pneumatic power transmission systems. The only difference was the stuff being delivered was electricity not gas, and the "pipes" doing the delivering were flexible.

The California Electric Company (now PG&E) in San Francisco in 1879 used two direct current generators from Charles Brush's company to supply multiple customers with power for their arc lights. This San Francisco system was the first case of a utility selling electricity from a central plant to multiple customers via transmission lines. CEC soon opened a second plant with 4 additional generators. Service charges for light from sundown to midnight was $10 per lamp per week.

In December 1880, Brush Electric Company set up a central station to supply a 2-mile (3.2 km) length of Broadway with arc lighting. By the end of 1881, New York, Boston, Philadelphia, Baltimore, Montreal, Buffalo, San Francisco, Cleveland and other cities had Brush arc light systems, producing public light well into the 20th century. By 1893 there were 1500 arc lights illuminating New York streets.

The first electricity system supplying incandescent lights was built by Edison Electric Illuminating Company in lower Manhattan eventually serving one square mile with 6 "jumbo dynamos" housed at Pearl Street Station. When service began in September 1882, there were 85 customers with 400 light bulbs. Each dynamo produced 100 kW- enough for 1200 incandescent lights, and transmission was at 110 volts via underground conduits. The system cost $300,000 to build with installation of the 100,000 feet (30,000 m) of underground conduits one of the most expensive parts of the project. Operating expenses exceeded income in the first two years and fire destroyed the plant in 1890. Edison's lights were cheaper, provided light that was warmer and operated at much lower voltages than the arc lights. Further, Edison had a three wire system so that either 110 volts or 220 volts could be supplied to power some motors.

Availability of Large Scale Generation

Availability of large amounts of power from diverse locations would become possible after Charles Parsons' production of turbogenerators beginning 1889. Turbogenerator output quickly jumped from 100 kW to 25 megawatts in two decades. Prior to efficient turbogenerators, hydroelectric projects were a significant source of large amounts of power requiring transmission infrastructure.

Induction Coils Advantage of Alternating Current

When George Westinghouse became interested in electricity, he quickly and correctly concluded that Edison's low voltages were too inefficient to be scaled up for transmission needed for large systems. He further understood that long-distance transmission needed high voltage and that inexpensive conversion technology only existed for alternating current. Transformers would play the decisive role in the victory of alternating current over direct current for transmission and distribution systems. In 1876, Pavel Yablochkov patented his mechanism of using induction coils to serve as a step up transformer prior to the Paris Exposition demonstrating his arc lights. Lucien Gaulard and John Dixon Gibbs later developed more efficient, less expensive AC transformers.

The Birth of the First Transformer

Between 1884 and 1885, Hungarian engineers Zipernowsky, Bláthy and Déri from the Ganz company in Budapest created the efficient "Z.B.D." closed-core coils, as well as the modern electric distribution system. The three had discovered that all former coreless or open-core devices were incapable of regulating voltage, and were therefore impracticable. Their joint patent described two versions of a design with no poles: the "closed-core transformer" and the "shell-core transformer". Ottó Bláthy suggested the use of closed-cores, Károly Zipernowsky the use of shunt connections, and Miksa Déri performed the experiments.

In the closed-core transformer the iron core is a closed ring around which the two coils are wound. In the shell type transformer, the windings are passed through the core. In both designs, the magnetic flux linking the primary and secondary windings travels almost entirely within the iron core, with no intentional path through air. The core consists of iron strands or sheets. These revolutionary design elements would finally make it technically and economically feasible to provide electric power for lighting in homes, businesses and public spaces. Zipernowsky, Bláthy and Déri also discovered the transformer formula, Vs/Vp = Ns/Np. Electrical and electronic systems the world over rely on the principles of the original Ganz transformers. The inventors are also credited with the first use of the word "transformer" to describe a device for altering the EMF of an electric current.

The concept that is the basis of modern transmission using inexpensive step up and step down transformers was first implemented by Westinghouse, Stanley and Franklin Leonard Pope in 1886 in Great Barrington, Massachusetts. There were still problems with efficient generators and high voltage transformers. At an AIEE meeting on May 16, 1888, Nikola Tesla delivered a lecture entitled *A New System of Alternating Current Motors and Transformers*, describing the equipment which allowed efficient generation and use of alternating currents. Westinghouse needed Tesla's better step up transformer technology and bought patents for it along with the highly efficient and inexpensive polyphase design for AC generators and motors used today. The utter simplicity of polyphase generators and motors meant that besides their efficiency they could be manufactured cheaply, compactly and would required little attention to maintain. Simple economics would drive the expensive, balky and mechanically complex DC dynamos to their ultimate extinction. As it turned out, the deciding factor in the War of Currents was the availability of low cost step up and step down transformers that meant that all customers regardless of their specialised voltage requirements could be served at minimal cost of conversion. This "universal system" is today regarded as one of the most influential innovations for the use of electricity.

High Voltage Direct Current Transmission

The case for alternating current was not clear at the turn of the century and high voltage direct current transmission systems were successfully installed without the benefit of transformers. Rene Thury who had spent six months at Edison's Menlo park facility understood his problem with transmission and was convinced that moving electricity over great distances was possible using direct current. He was familiar with the work of Marcel Deprez, who did early work on high voltage transmission after being inspired by the capability of arc light generators to support lights over great distances. Deprez avoided transformers by placing generators and loads in series as arc light systems of Charles F. Brush did. Thury developed this idea into the first commercial system for high-voltage DC transmission. Like Brush's dynamos, current is kept constant, and when increasing load demands more pressure, voltage is increased.

The Thury System was successfully used on several DC transmission projects from Hydro generators. The first in 1885 was a low voltage system in Bözingen, and the first high voltage system went into service in 1889 in Genoa, Italy by the *Acquedotto de Ferrari-Galliera* company. This system transmitted 630 kW at 14 kV DC over a circuit 120 km long. The largest Thury System was the Lyon Moutiers project that was 230 km in length, eventually delivering 20 Megawatts, at 125kV.

Victory for AC

Ultimately, the versatility of the Thury system was hampered the fragility of series distribution, and the lack of a reliable DC conversion technology that would not show up until the 1940s with improvements in mercury arc valves. The AC "universal system" won by force of numbers, proliferating systems with transformers both to couple generators to high-voltage transmission lines, and to connect transmission to local distribution circuits. By a suitable choice of utility frequency, both lighting and motor loads could be served. Rotary converters and later mercury-arc valves and other rectifier equipment allowed DC load to be served by local conversion where needed. Even generating stations and loads using different frequencies could also be interconnected using rotary converters. By using common generating plants for every type of load, important economies of scale were achieved, lower overall capital investment was required, load factor on each plant was increased allowing for higher efficiency, allowing for a lower cost of energy to the consumer and increased overall use of electric power.

By allowing multiple generating plants to be interconnected over a wide area, electricity production cost was reduced. The most efficient available plants could be used to supply the varying loads during the day. Reliability was improved and capital investment cost was reduced, since stand-by generating capacity could be shared over many more customers and a wider geographic area. Remote and low-cost sources of energy, such as hydroelectric power or mine-mouth coal, could be exploited to lower energy production cost.

The first transmission of three-phase alternating current using high voltage took place in 1891 during the international electricity exhibition in Frankfurt. A 25 kV transmission line, approximately 175 kilometres long, connected Lauffen on the Neckar and Frankfurt.

Initially transmission lines were supported by porcelain pin-and-sleeve insulators similar to those used for telegraphs and telephone lines. However, these had a practical limit of 40 kV. In 1907, the invention of the disc insulator by Harold W. Buck of the Niagara Falls Power Corporation and Edward M. Hewlett of General Electric allowed practical insulators of any length to be constructed for higher voltages.

The first large scale hydroelectric generators in the USA were installed in 1895 at Niagara Falls and provided electricity to Buffalo, New York via power transmission lines. A statue of Tesla stands at Niagara Falls today in tribute to his contributions.

Modern Period

Voltages used for electric power transmission increased throughout the 20th century. The first electric power transmission line in North America operated at 4000V. It was constructed in 1889 between the generating station at Willamette Falls in Oregon City, Oregon, and downtown Portland, Oregon, a distance of around 13 miles. By 1914 fifty-five transmission systems operating at more than 70,000 V were in service, and the highest voltage then used was 150,000 volts. The first three-phase alternating current power transmission at 110 kV took place in 1907 between Croton and Grand Rapids, Michigan. Voltages of 100 kV and more where not established technology until around 5 years later, with for example the first 110 kV line in Europe between Lauchhammer and Riesa, Germany in 1912.

In the early 1920s the Pit River – Cottonwood – Vaca-Dixon line was built for 220 kV transporting power from hydroelectric plants in the Sierra Nevada to the San Francisco Bay Area, at the same time the Big Creek-Los Angeles lines where upgraded to the same voltage. Both of those systems entered commercial service in 1923. On April 17, 1929 the first 220 kV line in Germany was completed, running from Brauweiler near Cologne, over Kelsterbach near Frankfurt,

Rheinau near Mannheim, Ludwigsburg–Hoheneck near Austria. This line comprises the North-South interconnect, at the time one of the world's largest power systems. The masts of this line were designed for eventual upgrade to 380 kV. However the first transmission at 380 kV in Germany was on October 5, 1957 between the substations in Rommerskirchen and Ludwigsburg–Hoheneck.

The world's first 380 kV power line was built in Sweden, the 952 km Harsprånget-Hallsberg line in 1952. In 1965, the first extra-high-voltage transmission at 735 kV took place on a Hydro-Québec transmission line. In 1982 the first transmission at 1200 kV was in the Soviet Union. The rapid industrialisation in the 20th century made electrical transmission lines and grids a critical part of the economic infrastructure in most industrialised nations. Interconnection of local generation plants and small distribution networks was greatly spurred by the requirements of World War I, where large electrical generating plants were built by governments to provide power to munitions factories; later these plants were connected to supply civil load through long-distance transmission.

Small municipal electrical utilities did not necessarily desire to reduce the cost of each unit of electricity sold; to some extent, especially during the period 1880–1890, electrical lighting was considered a luxury product and electric power was not substituted for steam power. Engineers such as Samuel Insull in the United States and Sebastian Z. De Ferranti in the United Kingdom were instrumental in overcoming technical, economic, regulatory and political difficulties in development of long-distance electric power transmission. By introduction of electric power transmission networks, in the city of London the cost of a kilowatt-hour was reduced to one-third in a ten-year period.

In 1926 electrical networks in the United Kingdom began to be interconnected in the National Grid, initially operating at 132,000 volts.

Bulk Power Transmission

Engineers design transmission networks to transport the energy as efficiently as feasible, while at the same time taking into account economic factors, network safety and redundancy. These networks use components such as power lines, cables, circuit breakers, switches and transformers. The transmission network is usually administered on a regional basis by an entity such as a regional transmission organisation or transmission system operator.

Figure : A transmission substation decreases the voltage of incoming electricity, allowing it to connect from long distance high voltage transmission, to local lower voltage distribution. It also reroutes power to other transmission lines that serve local markets. This is the PacifiCorp Hale Substation, Orem, Utah, USA.

Transmission efficiency is hugely improved by devices that increase the voltage, and proportionately reduce the current in the conductors, thus keeping the power transmitted nearly equal to the power input. The reduced current flowing through the line reduces the losses in the conductors. According to Joule's Law, energy losses are directly proportional to the square of the current. Thus, reducing the current by a factor of 2 will lower the energy lost to conductor resistance by a factor of 4.

This change in voltage is usually achieved in AC circuits using a *step-up transformer*. DC systems require relatively costly conversion equipment which may be economically justified for particular projects, but are less common currently.

A transmission grid is a network of power stations, transmission circuits, and substations. Energy is usually transmitted within a grid with three-phase AC. Single phase AC is used only for distribution to end users since it is not usable for large polyphase induction motors. In the 19th century, two-phase transmission was used but required either three wires with unequal currents or four wires. Higher order phase systems require more than three wires, but deliver marginal benefits.

The capital cost of electric power stations is so high, and electric demand is so variable, that it is often cheaper to import some portion of the needed power than to generate it locally. Because nearby loads are often correlated (hot weather in the Southwest portion of the US might cause many people to use air conditioners), electricity often comes from distant sources. Because of the economics of load balancing, wide area transmission grids now span across countries and even large portions of continents. The web of interconnections between power producers and consumers ensures that power can flow, even if a few links are inoperative.

The unvarying (or slowly varying over many hours) portion of the electric demand is known as the *base load* and is generally served best by large facilities (which are therefore efficient due to economies of scale) with low variable costs for fuel and operations. Such facilities might be nuclear or coal-fired power stations, or hydroelectric, while other renewable energy sources such as concentrated solar thermal and geothermal power have the potential to provide base load power. Renewable energy sources such as solar photovoltaics, wind, wave, and tidal are, due to their intermittency, not considered "base load" but can still add power to the grid. The remaining power demand, if any, is supplied by peaking power plants, which are typically smaller, faster-responding, and higher cost sources, such as combined cycle or combustion turbine plants fuelled by natural gas.

Long-distance transmission of electricity (thousands of kilometers) is cheap and efficient, with costs of US$0.005–0.02/kWh (compared to annual averaged large producer costs of US$0.01–0.025/kWh, retail rates upwards of US$0.10/kWh, and multiples of retail for instantaneous suppliers at unpredicted highest demand moments). Thus distant suppliers can be cheaper than local sources (e.g., New York City buys a lot of electricity from Canada). Multiple local sources (even if more expensive and infrequently used) can make the transmission grid more fault tolerant to weather and other disasters that can disconnect distant suppliers.

Long distance transmission allows remote renewable energy resources to be used to displace fossil fuel consumption. Hydro and wind sources cannot be moved closer to populous cities, and solar costs are lowest in remote areas where local power needs are minimal.

Connection costs alone can determine whether any particular renewable alternative is economically sensible. Costs can be prohibitive for transmission lines, but various proposals for massive infrastructure investment in high capacity, very long distance super grid transmission networks could be recovered with modest usage fees.

Grid Input

At the generating plants the energy is produced at a relatively low voltage between about 2.3 kV and 30 kV, depending on the size of the unit. The generator terminal voltage is then stepped up by the power station transformer to a higher voltage (115 kV to 765 kV AC, varying by the transmission system and by country) for transmission over long distances.

Losses

Transmitting electricity at high voltage reduces the fraction of energy lost to resistance, which averages around 7%. For a given amount of power, a higher voltage reduces the current and thus the resistive losses in the conductor. For example, raising the voltage by a factor of 10 reduces the current by a corresponding factor of 10 and therefore the I^2R losses by a factor of 100, provided the same sized conductors are used in both cases. Even if the conductor size is reduced 10-fold to match the lower current the I^2R losses are still reduced 10-fold. Long distance transmission is typically done with overhead lines at voltages of 115 to 1,200 kV. At extremely high voltages, more than 2,000 kV between conductor and ground, corona discharge losses are so large that they can offset the lower resistance loss in the line conductors. Measures to reduce corona losses include conductors having large diameter; often hollow to save weight, or bundles of two or more conductors. Transmission and distribution losses in the USA were estimated at 6.6% in 1997 and 6.5% in 2007. In general, losses are estimated from the discrepancy between energy produced (as reported by power plants) and energy sold to end customers; the difference between what is produced and what is consumed constitute transmission and distribution losses. As of 1980, the longest cost-effective distance for DC electricity was determined to be 7,000 km (4,300 mi). For AC it was 4,000 km (2,500 mi), though all transmission lines in use today are substantially shorter.

In an alternating current circuit, the inductance and capacitance of the phase conductors can be significant. The currents that flow in these components of the circuit impedance constitute reactive power, which transmits no energy to the load. Reactive current causes extra losses in the transmission circuit. The ratio of real power (transmitted to the load) to apparent power is the power factor. As reactive current increases, the reactive power increases and the power factor decreases. For systems with low power factors, losses are higher than for systems with high power factors. Utilities add capacitor banks and other components (such as phase-shifting transformers; static VAR compensators; physical transposition of the phase conductors; and flexible AC transmission systems, FACTS) throughout the system to control reactive power flow for reduction of losses and stabilisation of system voltage.

Subtransmission

Subtransmission is part of an electric power transmission system that runs at relatively lower voltages. It is uneconomical to connect all distribution substations to the high main transmission voltage, because the equipment is larger and more expensive. Typically, only larger substations connect with this high voltage. It is stepped down and sent to smaller substations in towns and neighbourhoods. Subtransmission circuits are usually arranged in loops so that a single line failure does not cut off service to a large number of customers for more than a short time. While subtransmission circuits are usually carried on overhead lines, in urban areas buried cable may be used.

There is no fixed cutoff between subtransmission and transmission, or subtransmission and distribution. The voltage ranges overlap somewhat. Voltages of 69 kV, 115 kV and 138 kV are often used for subtransmission in North America. As power systems evolved, voltages formerly used for transmission were used for subtransmission, and subtransmission voltages became distribution voltages. Like transmission, subtransmission moves relatively large amounts of power, and like distribution, subtransmission covers an area instead of just point to point.

Transmission Grid Exit

At the substations, transformers reduce the voltage to a lower level for distribution to commercial and residential users. This distribution is accomplished with a combination of sub-transmission (33 kV to 132 kV) and distribution (3.3 to 25 kV). Finally, at the point of use, the energy is transformed to low voltage.

High-voltage Direct Current

A high-voltage, direct current (HVDC) electric power transmission system uses direct current for the bulk transmission of electrical power, in contrast with the more common alternating current systems. For long-distance transmission, HVDC systems may be less expensive and suffer lower electrical losses. For underwater power cables, HVDC avoids the heavy currents required by the cable capacitance. For shorter distances, the higher cost of DC conversion equipment compared to an AC system may still be warranted, due to other benefits of direct current links. HVDC allows power transmission between unsynchronised AC distribution systems, and can increase system stability by preventing cascading failures from propagating from one part of a wider power transmission grid to another.

Figure : Long distance HVDC lines carrying hydroelectricity from Canada's Nelson river to this station where it is converted to AC for use in Winnipeg's local grid

The modern form of HVDC transmission uses technology developed extensively in the 1930s in Sweden at ASEA. Early commercial installations included one in the Soviet Union in 1951 between Moscow and Kashira, and a 10–20 MW system between Gotland and mainland Sweden in 1954. The longest HVDC link in the world is currently the Xiangjiaba-Shanghai 2,071 km (1,287 mi) 6400 MW link connecting the Xiangjiaba Dam to Shanghai, in the People's Republic of China. In 2012, the longest HVDC link will be the Rio Madeira link connecting the Amazonas to the São Paulo area where the length of the DC line is over 2,500 km (1,600 mi).

High Voltage Transmission

High voltage (in either AC or DC electrical power transmission applications) is used for electric power transmission to reduce the energy lost in the resistance of the wires. For a given quantity of power transmitted and size of conductor, doubling the voltage will deliver the same power at only half the current. Since the power lost as heat in the wires is proportional to the square of the current, but does not depend in any major way on the voltage delivered by the power line, doubling the voltage in a power system reduces the line-loss loss per unit of electrical power delivered by a factor of 4. Power loss in transmission lines can also be reduced by reducing resistance, for example by increasing the diameter of the conductor; but larger conductors are heavier and more expensive.

High voltages cannot easily be used for lighting and motors, and so transmission-level voltages must be reduced to values compatible with end-use equipment. Transformers are used to change the voltage level in alternating current (AC) transmission circuits. The competition between the direct current (DC) of Thomas Edison and the AC of Nikola Tesla and George Westinghouse was known as the War of Currents, with AC becoming dominant. Practical manipulation of high power high voltage DC became possible with the development of high power electronic rectifier devices such as mercury arc valves and, more recently starting in the 1970s, high power semiconductor devices such as high power thyristors and 21st century high power variants such as integrated gate-commutated thyristors (IGCTs), MOS controlled thyristors (MCTs) and insulated-gate bipolar transistors (IGBT).

History of HVDC Transmission

Electromechanical (Thury) Systems

The first long-distance transmission of electric power was demonstrated using direct current in 1882 at the Miesbach-Munich Power Transmission, but only 2.5 kW was transmitted. An early method of high-voltage DC transmission was developed by the Swiss engineer René Thury and his method was put into practice by 1889 in Italy by the *Acquedotto De Ferrari-Galliera* company. This system used series-connected motor-generator sets to increase voltage. Each set was insulated from ground and driven by insulated shafts from a prime mover. The line was operated in constant current mode, with up to 5,000 volts on each machine, some machines having double commutators to reduce the voltage on each commutator. This system transmitted 630 kW at 14 kV DC over a distance of 120 km. The Moutiers-Lyon system transmitted 8,600 kW of hydroelectric power a distance of 200 km, including 10 km of underground cable. The system used eight series-connected generators with dual commutators for a total voltage of 150,000 volts between the poles, and ran from about 1906 until 1936. Fifteen Thury systems were in operation by 1913 Other Thury systems operating at up to 100 kV DC operated up to the 1930s, but the rotating machinery required high maintenance and had high energy loss. Various other electromechanical devices were tested during the first half of the 20th century with little commercial success.

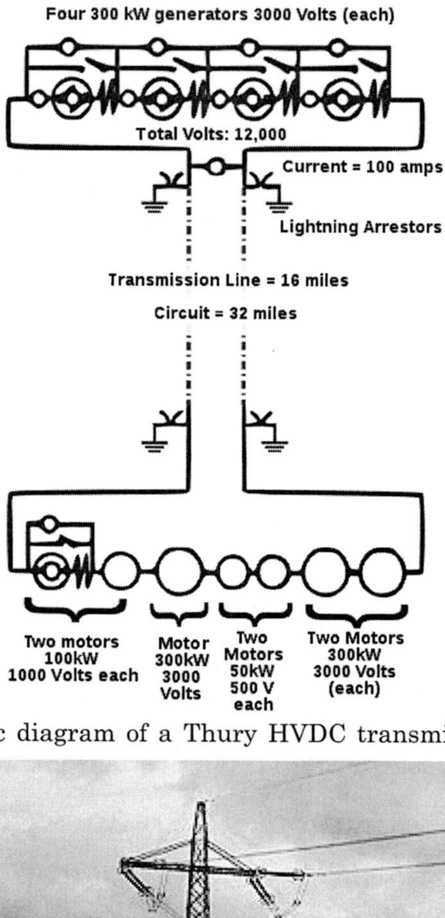

Figure : Schematic diagram of a Thury HVDC transmission system

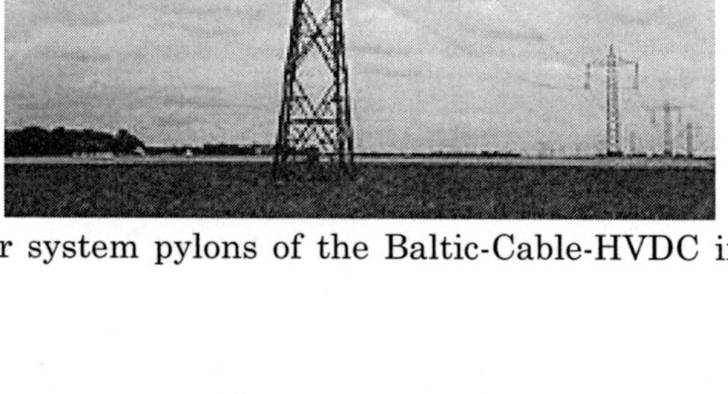

Figure : Bipolar system pylons of the Baltic-Cable-HVDC in Sweden

Figure : HVDC in 1971: this 150 kV mercury arc valve converted AC hydropower voltage for transmission to distant cities from Manitoba Hydro generators.

One conversion technique attempted for conversion of direct current from a high transmission voltage to lower utilisation voltage was to charge series-connected batteries, then connect the batteries in parallel to serve distribution loads. While at least two commercial installations were tried around the turn of the 20th century, the technique was not generally useful owing to the limited capacity of batteries, difficulties in switching between series and parallel connections, and the inherent energy inefficiency of a battery charge/discharge cycle.

Mercury Arc Valves

The grid controlled mercury arc valve became available for power transmission during the period 1920 to 1940. Starting in 1932, General Electric tested mercury-vapour valves and a 12 kV DC transmission line, which also served to convert 40 Hz generation to serve 60 Hz loads, at Mechanicville, New York. In 1941, a 60 MW, +/-200 kV, 115 km buried cable link was designed for the city of Berlin using mercury arc valves (Elbe-Project), but owing to the collapse of the German government in 1945 the project was never completed. The nominal justification for the project was that, during wartime, a buried cable would be less conspicuous as a bombing target. The equipment was moved to the Soviet Union and was put into service there. Introduction of the fully static mercury arc valve to commercial service in 1954 marked the beginning of the modern era of HVDC transmission. A HVDC-connection was constructed by ASEA between the mainland of Sweden and the island Gotland.

Mercury arc valves require an external circuit to force the current to zero and thus turn off the valve. In HVDC applications, the AC power system itself provides the means of "commutating" the current to another valve in the converter. Consequently, converters built with mercury arc valves are known as line-commutated converters (LCC). Line-commutated converters require rotating synchronous machines in the AC systems to which they are connected, making power transmission into a passive load impossible. Mercury arc valves were common in systems designed up to 1975, the last mercury arc HVDC system (the Nelson River Bipole 1 system in Manitoba, Canada) having been put into service in 1976. As of 2011 the Inter-Island HVDC (high voltage direct current) link between the North and South Islands of New Zealand is the last major operating mercury arc HVDC not yet replaced with a solid state system (this is being planned for 2012).

Thyristor Valves

Since 1976, new HVDC systems have used only solid-state devices, in most cases thyristor valves. Like mercury arc valves, thyristors require an external circuit (in HVDC applications, the AC power system itself does this) to turn them off, and HVDC using thyristor valves is therefore also known as line-commutated converter (LCC) HVDC and has the same restrictions. On March 15, 1979, a thyristor based direct current connection between Cabora Bassa and Johannesburg (1410 km, ±533 kV, 1920 MW) was turned on. Though the electronics were built in 1974 by AEG, and BBC (Brown Boveri Company) and Siemens were partners in the project, the late turn on was a result of the civil war.

Capacitor-Commutated Converters (CCC)

Line-Commutated Converters have some limitations in their use for HVDC systems, resulting from the inability of the thyristor to turn off current and its need for a period of reverse voltage after turn-off (turn-off time). An attempt to address these limitations is the "Capacitor-Commutated Converter (CCC)" which has been used in a small number of HVDC systems. The CCC is a conventional, Line-Commutated Converter HVDC system using thyristor valves, in which series capacitors are inserted into the AC line connections, either on the primary or secondary side of the converter transformer. The series capacitors partially offset the "commutating inductance" of the converter and help to reduce fault currents. This allows a smaller "extinction angle" to be used at an inverter, reducing the requirement for reactive power. However, CCC has remained only a niche application of HVDC because of the advent of Voltage-Sourced Converter (VSC) HVDC which completely eliminates the need for an extinction time.

The Future

According to Sood, the next 25 years may well be dominated by "force commutated converters" (by which he means the above-mentioned capacitor commutated converters (CCC) and also thyristor or thyristor-like semiconductors with gates that are actively controlled by separate switching circuitry, for smoother switching response). However, the force-commutated converters using capacitative commutation never outgrew niche use, due to the rapid emergence of the second trend today, in favour of "self-commutated converters" which use more complicated semiconductor devices which can be directly turned off by gate action. These self-commutated devices include the insulated gate bipolar transistors (IGBT) and variants of the thyristor such as the integrated gate-commutated thyristor (IGCT), and gate turn-off thyristor (GTO). Self-commutated converters, more commonly known as "Voltage-Sourced Converters" (or "Voltage-Source Converters") started to appear in 1997 with the experimental Hellsjön–Grängesberg project in Sweden, and by the end of 2011, this technology had captured a significant proportion of the HVDC market. VSC HVDC systems will probably eventually replace all simple thyristor-based systems now in use, even for very highest power transmission DC applications.

Advantages of HVDC over AC transmission

The advantage of HVDC is the ability to transmit large amounts of power over long distances with lower capital costs and with lower losses than AC. Depending on voltage level and construction details, losses are quoted as about 3% per 1,000 km. High-voltage direct current transmission allows efficient use of energy sources, remote from load centres. In a number of applications HVDC is more effective than AC transmission. Examples include:

- Undersea cables, where high capacitance causes additional AC losses. (e.g., 250 km Baltic Cable between Sweden and Germany, the 600 km NorNed cable between Norway and the Netherlands, and 290 km Basslink between the Australian mainland and Tasmania)

- Endpoint-to-endpoint long-haul bulk power transmission without intermediate 'taps', for example, in remote areas
- Increasing the capacity of an existing power grid in situations where additional wires are difficult or expensive to install
- Power transmission and stabilisation between unsynchronised AC distribution systems
- Connecting a remote generating plant to the distribution grid, for example the Nelson River DC Transmission System
- Stabilising a predominantly AC power-grid, without increasing prospective short circuit current
- Reducing line cost. HVDC needs fewer conductors as there is no need to support multiple phases. Also, thinner conductors can be used since HVDC does not suffer from the skin effect
- Facilitate power transmission between different countries that use AC at differing frequencies
- Synchronise AC produced by renewable energy sources.

Long undersea / underground high voltage cables have a high electrical capacitance, since the conductors are surrounded by a relatively thin layer of insulation and a metal sheath while the extensive length of the cable multiplies the area between the conductors. The geometry is that of a long co-axial capacitor. Where alternating current is used for cable transmission, this capacitance appears in parallel with load. Additional current must flow in the cable to charge the cable capacitance, which generates additional losses in the conductors of the cable. Additionally, there is a dielectric loss component in the material of the cable insulation, which consumes power.

However, when direct current is used, the cable capacitance is charged only when the cable is first energised or when the voltage is changed; there is no steady-state additional current required. For a long AC undersea cable, the entire current-carrying capacity of the conductor could be used to supply the charging current alone. The cable capacitance issue limits the length and power carrying capacity of AC cables. DC cables have no such limitation, and are essentially bound by only Ohm's Law. Although some DC leakage current continues to flow through the dielectric insulators, this is very small compared to the cable rating and much less than with AC transmission cables.

HVDC can carry more power per conductor because, for a given power rating, the constant voltage in a DC line is the same as the peak voltage in an AC line. The power delivered in an AC system is defined by the root mean square (RMS) of an AC voltage, but RMS is only about 71% of the peak voltage. The peak voltage of AC determines the actual insulation thickness and conductor spacing. Because DC operates at a constant maximum voltage, this allows existing transmission line corridors with equally sized conductors and insulation to carry more power into an area of high power consumption than AC, which can lower costs.

Because HVDC allows power transmission between unsynchronised AC distribution systems, it can help increase system stability, by preventing cascading failures from propagating from one part of a wider power transmission grid to another. Changes in load that would cause portions of an AC network to become unsynchronised and separate would not similarly affect a DC link, and the power flow through the DC link would tend to stabilise the AC network. The magnitude and direction of power flow through a DC link can be directly commanded, and changed as needed to support the AC networks at either end of the DC link. This has caused many power system operators to contemplate wider use of HVDC technology for its stability benefits alone.

Disadvantages

The disadvantages of HVDC are in conversion, switching, control, availability and maintenance. HVDC is less reliable and has lower availability than alternating current (AC) systems, mainly due to the extra conversion equipment. Single-pole systems have availability of about 98.5%, with about a third of the downtime unscheduled due to faults. Fault-tolerant bipole systems provide high availability for 50% of the link capacity, but availability of the full capacity is about 97% to 98%.

The required static inverters are expensive and have limited overload capacity. At smaller transmission distances, the losses in the static inverters may be bigger than in an AC transmission line. The cost of the inverters may not be offset by reductions in line construction cost and lower line loss.

In contrast to AC systems, realising multiterminal systems is complex, as is expanding existing schemes to multiterminal systems. Controlling power flow in a multiterminal DC system requires good communication between all the terminals; power flow must be actively regulated by the inverter control system instead of the inherent impedance and phase angle properties of the transmission line. Multi-terminal lines are rare. One is in operation at the Hydro Québec –

New England transmission from Radisson to Sandy Pond. Another example is the Sardinia-mainland Italy link which was modified in 1989 to also provide power to the island of Corsica.

HVDC circuit breakers are difficult to build because some mechanism must be included in the circuit breaker to force current to zero, otherwise arcing and contact wear would be too great to allow reliable switching. Operating a HVDC scheme requires many spare parts to be kept, often exclusively for one system, as HVDC systems are less standardised than AC systems and technology changes faster.

Costs of High Voltage DC Transmission

Normally manufacturers such as Alstom, Siemens and ABB do not state specific cost information of a particular project since this is a commercial matter between the manufacturer and the client.

Costs vary widely depending on the specifics of the project such as power rating, circuit length, overhead vs. underwater route, land costs, and AC network improvements required at either terminal. A detailed evaluation of DC vs. AC cost may be required where there is no clear technical advantage to DC alone and only economics drives the selection.

However some practitioners have given out some information that can be reasonably well relied upon:

For an 8 GW 40 km link laid under the English Channel, the following are approximate primary equipment costs for a 2000 MW 500 kV bipolar conventional HVDC link (exclude way-leaving, on-shore reinforcement works, consenting, engineering, insurance, etc.)

- Converter stations ~£110M
- Subsea cable + installation ~£1M/km

So for an 8 GW capacity between England and France in four links, little is left over from £750M for the installed works. Add another £200–300M for the other works depending on additional onshore works required.

An April, 2010 announcement for a 2,000 MW line, 64 km, between Spain and France, is 700 million euros; this includes the cost of a tunnel through the Pyrenees.

Rectifying and Inverting

Components

Most of the HVDC systems in operation today are based on *Line-Commutated Converters*. Early static systems used mercury arc rectifiers, which were unreliable and required high maintenance. Two HVDC systems using mercury arc rectifiers are still in service (as of 2008). The thyristor valve was first used in HVDC systems in the 1960s. The thyristor is a solid-state semiconductor device similar to the diode, but with an extra control terminal that is used to switch the device on at a particular instant during the AC cycle. The insulated-gate bipolar transistor (IGBT) is now also used, forming a *Voltage Sourced Converter*, and offers simpler control, reduced harmonics and reduced valve cost.

Figure : Two of three thyristor valve stacks used for long distance transmission of power from Manitoba Hydro dams

Because the voltages in HVDC systems, up to 800 kV in some cases, exceed the breakdown voltages of the semiconductor devices, HVDC converters are built using large numbers of semiconductors in series.

The low-voltage control circuits used to switch the thyristors on and off need to be isolated from the high voltages present on the transmission lines. This is usually done optically. In a hybrid control system, the low-voltage control electronics sends light pulses along optical fibres to the *high-side* control electronics. Another system, called *direct light triggering*, dispenses with most of the high-side electronics, instead using light pulses from the control electronics to switch light-triggered thyristors (LTTs), although a small monitoring electronics unit may still be required for protection of the valve.

A complete switching element is commonly referred to as a *valve*, irrespective of its construction.

Rectifying and Inverting Systems

Rectification and inversion use essentially the same machinery. Many substations (*Converter Stations*) are set up in such a way that they can act as both rectifiers and inverters. At the AC end a set of transformers, often three physically separated single-phase transformers, isolate the station from the AC supply, to provide a local earth, and to ensure the correct eventual DC voltage. The output of these transformers is then connected to a bridge rectifier formed by a number of valves. The basic configuration uses six valves, connecting each of the three phases to each of the two DC rails. However, with a phase change only every sixty degrees, considerable harmonics remain on the DC rails. An enhancement of this configuration uses 12 valves (often known as a twelve-pulse system). The AC is split into two separate three phase supplies before transformation. One of the sets of supplies is then configured to have a star (wye) secondary, the other a delta secondary, establishing a thirty degree phase difference between the two sets of three phases. With twelve valves connecting each of the two sets of three phases to the two DC rails, there is a phase change every 30 degrees, and harmonics are considerably reduced.

In addition to the conversion transformers and valve-sets, various passive resistive and reactive components help filter harmonics out of the DC rails.

Configurations

Monopole and Earth Return

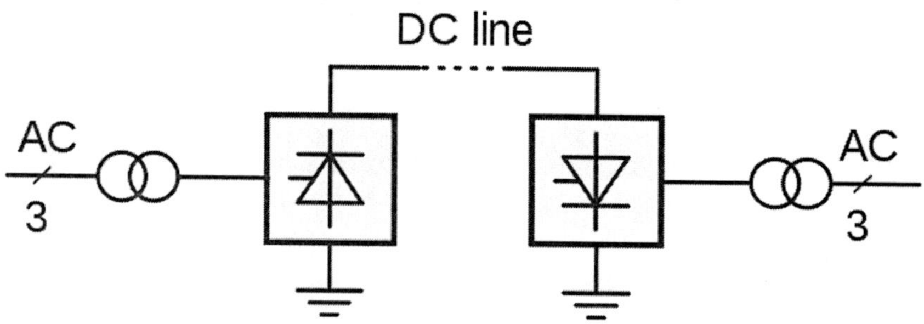

Figure : Block diagram of a monopole system with earth return

In a common configuration, called monopole, one of the terminals of the rectifier is connected to earth ground. The other terminal, at a potential high above or below ground, is connected to a transmission line. The earthed terminal may be connected to the corresponding connection at the inverting station by means of a second conductor. If no metallic conductor is installed, current flows in the earth between the earth electrodes at the two stations. Therefore it is a type of single wire earth return. The issues surrounding earth-return current include:

- Electrochemical corrosion of long buried metal objects such as pipelines
- Underwater earth-return electrodes in seawater may produce chlorine or otherwise affect water chemistry.
- An unbalanced current path may result in a net magnetic field, which can affect magnetic navigational compasses for ships passing over an underwater cable.

These effects can be eliminated with installation of a metallic return conductor between the two ends of the monopolar transmission line. Since one terminal of the converters is connected to earth, the return conductor need not be insulated for the full transmission voltage which makes it less costly than the high-voltage conductor. Use of a metallic return conductor is decided based on economic, technical and environmental factors. Modern monopolar systems for pure overhead lines carry typically 1,500 MW. If underground or underwater cables are used, the typical value is 600 MW. Most monopolar systems are designed for future bipolar expansion. Transmission line towers may be designed to carry two conductors, even if only one is used initially for the monopole transmission system. The second conductor is either unused, used as electrode line or connected in parallel with the other (as in case of Baltic-Cable).

Bipolar

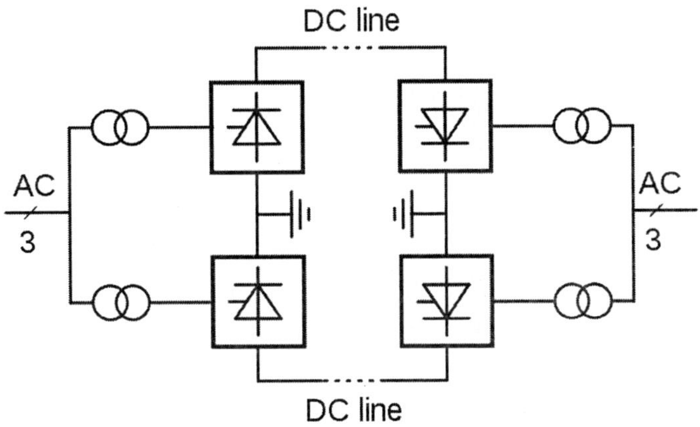

Figure : Block diagram of a bipolar system that also has an earth return

In bipolar transmission a pair of conductors is used, each at a high potential with respect to ground, in opposite polarity. Since these conductors must be insulated for the full voltage, transmission line cost is higher than a monopole with a return conductor. However, there are a number of advantages to bipolar transmission which can make it the attractive option.

- Under normal load, negligible earth-current flows, as in the case of monopolar transmission with a metallic earth-return. This reduces earth return loss and environmental effects.
- When a fault develops in a line, with earth return electrodes installed at each end of the line, approximately half the rated power can continue to flow using the earth as a return path, operating in monopolar mode.
- Since for a given total power rating each conductor of a bipolar line carries only half the current of monopolar lines, the cost of the second conductor is reduced compared to a monopolar line of the same rating.
- In very adverse terrain, the second conductor may be carried on an independent set of transmission towers, so that some power may continue to be transmitted even if one line is damaged.

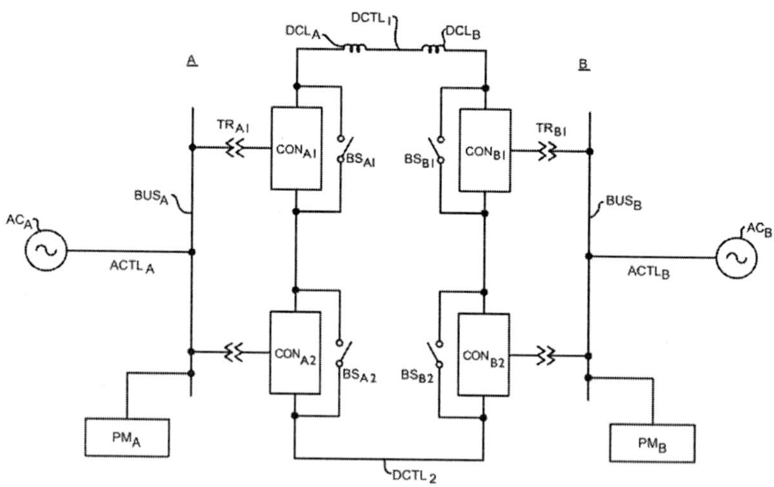

Figure : A block diagram of a bipolar HVDC transmission system, between two stations designated A and B. AC – represents an alternating current network CON – represents a converter valve, either rectifier or inverter, TR represents a power transformer, DCTL is the direct-current transmission line conductor.

A bipolar system may also be installed with a metallic earth return conductor. Bipolar systems may carry as much as 3,200 MW at voltages of +/-600 kV. Submarine cable installations initially commissioned as a monopole may be upgraded with additional cables and operated as a bipole. A bipolar scheme can be implemented so that the polarity of one or both poles can be changed. This allows the operation as two parallel monopoles. If one conductor fails, transmission can still continue at reduced capacity. Losses may increase if ground electrodes and lines are not designed for the extra current in this mode. To reduce losses in this case, intermediate switching stations may be installed, at which line segments can be switched off or parallelised. This was done at Inga–Shaba HVDC.

Back to Back

A back-to-back station (or B2B for short) is a plant in which both static inverters and rectifiers are in the same area, usually in the same building. The length of the direct current line is kept as short as possible. HVDC back-to-back stations are used for

- coupling of electricity mains of different frequency (as in Japan; and the GCC interconnection between UAE [50 Hz] and Saudi Arabia [60 Hz] under construction in ±2009–2011)
- coupling two networks of the same nominal frequency but no fixed phase relationship (as until 1995/96 in Etzenricht, Dürnrohr, Vienna, and the Vyborg HVDC scheme).
- different frequency and phase number (for example, as a replacement for traction current converter plants).

The DC voltage in the intermediate circuit can be selected freely at HVDC back-to-back stations because of the short conductor length. The DC voltage is as low as possible, in order to build a small valve hall and to avoid series connections of valves. For this reason at HVDC back-to-back stations valves with the highest available current rating are used.

Systems with Transmission Lines

The most common configuration of an HVDC link is two inverter/ rectifier stations connected by an overhead power line. This is also a configuration commonly used in connecting unsynchronised grids, in long-haul power transmission, and in undersea cables.

Multi-terminal HVDC links, connecting more than two points, are rare. The configuration of multiple terminals can be series, parallel, or hybrid (a mixture of series and parallel). Parallel configuration tends to be used for large capacity stations, and series for lower capacity stations. An example is the 2,000 MW Quebec-New England Transmission system opened in 1992, which is currently the largest multi-terminal HVDC system in the world.

Tripole: Current-modulating Control

A scheme patented in 2004 (Current modulation of direct current transmission lines) is intended for conversion of existing AC transmission lines to HVDC. Two of the three circuit conductors are operated as a bipole. The third conductor is used as a parallel monopole, equipped with reversing valves (or parallel valves connected in reverse polarity). The parallel monopole periodically relieves current from one pole or the other, switching polarity over a span of several minutes. The bipole conductors would be loaded to either 1.37 or 0.37 of their thermal limit, with the parallel monopole always carrying +/- 1 times its thermal limit current. The combined RMS heating effect is as if each of the conductors is always carrying 1.0 of its rated current. This allows heavier currents to be carried by the bipole conductors, and full use of the installed third conductor for energy transmission. High currents can be circulated through the line conductors even when load demand is low, for removal of ice.

As of 2005, no tri-pole conversions are in operation, although a transmission line in India has been converted to bipole HVDC.

Cross-Skagerrak consists of 3 poles, from which 2 are switched in parallel and the third uses an opposite polarity with a higher transmission voltage. A similar arrangement is HVDC Inter-Island, but it consists of 2 parallel-switched inverters feeding in the same pole and a third one with opposite polarity and higher operation voltage.

Corona Discharge

Corona discharge is the creation of ions in a fluid (such as air) by the presence of a strong electric field. Electrons are torn from neutral air, and either the positive ions or the electrons are attracted to the conductor, while the charged particles drift. This effect can cause considerable power loss, create audible and radio-frequency interference, generate toxic compounds such as oxides of nitrogen and ozone, and bring forth arcing.

Both AC and DC transmission lines can generate coronas, in the former case in the form of oscillating particles, in the latter a constant wind. Due to the space charge formed around the conductors, an HVDC system may have about half the loss per unit length of a high voltage AC system carrying the same amount of power. With monopolar transmission the choice of polarity of the energised conductor leads to a degree of control over the corona discharge. In particular, the polarity of the ions emitted can be controlled, which may have an environmental impact on particulate condensation. (particles of different polarities have a different mean-free path.) Negative coronas generate considerably more ozone than positive coronas, and generate it further *downwind* of the power line, creating the potential for health effects. The use of a *positive* voltage will reduce the ozone impacts of monopole HVDC power lines.

Applications

Overview

The controllability of current-flow through HVDC rectifiers and inverters, their application in connecting unsynchronised networks, and their applications in efficient submarine cables mean that HVDC cables are often used at national boundaries for the exchange of power (in North America, HVDC connections divide much of Canada and the United States into several electrical regions that cross national borders, although the purpose of these connections is still to connect unsynchronised AC grids to each other). Offshore windfarms also require undersea cables, and their turbines are unsynchronised. In very long-distance connections between just two points, for example power transmission from a large hydroelectric power plant at a remote site to an urban area, it is of great interest and several schemes of these kind were built. For interconnections to Siberia, Canada, and the Scandinavian North, it may be as result of decreased line-costs of HVDC make also of interest, but however no such interconnection was realised as inverters are expensive.

AC Network Interconnections

AC transmission lines can interconnect only synchronised AC networks that oscillate at the same frequency and in phase. Many areas that wish to share power have unsynchronised networks. The power grids of the UK, Northern Europe and continental Europe are not united into a single synchronised network. Japan has 50 Hz and 60 Hz networks. Continental North America, while operating at 60 Hz throughout, is divided into regions which are unsynchronised: East, West, Texas, Quebec, and Alaska. Brazil and Paraguay, which share the enormous Itaipu Dam hydroelectric plant, operate on 60 Hz and 50 Hz respectively. However, HVDC systems make it possible to interconnect unsynchronised AC networks, and also add the possibility of controlling AC voltage and reactive power flow. A generator connected to a long AC transmission line may become unstable and fall out of synchronisation with a distant AC power system. An HVDC transmission link may make it economically feasible to use remote generation sites. Wind farms located off-shore may use HVDC systems to collect power from multiple unsynchronised generators for transmission to the shore by an underwater cable.

In general, however, an HVDC power line will interconnect two AC regions of the power-distribution grid. Machinery to convert between AC and DC power adds a considerable cost in power transmission. The conversion from AC to DC is known as rectification, and from DC to AC as inversion. Above a certain break-even distance (about 50 km for submarine cables, and perhaps 600–800 km for overhead cables), the lower cost of the HVDC electrical conductors outweighs the cost of the electronics.

The conversion electronics also present an opportunity to effectively manage the power grid by means of controlling the magnitude and direction of power flow. An additional advantage of the existence of HVDC links, therefore, is potential increased stability in the transmission grid.

Renewable Electricity Superhighways

Figure : Two HVDC lines cross near Wing, North Dakota.

A number of studies have highlighted the potential benefits of very wide area super grids based on HVDC since they can mitigate the effects of intermittency by averaging and smoothing the outputs of large numbers of geographically dispersed wind farms or solar farms. Czisch's study concludes that a grid covering the fringes of Europe could bring 100% renewable power (70% wind, 30% biomass) at close to today's prices. There has been debate over the technical feasibility of this proposal and the political risks involved in energy transmission across a large number of international borders. The construction of such green power superhighways is advocated in a white paper that was released by the American Wind Energy Association and the Solar Energy Industries Association

In January 2009, the European Commission proposed €300 million to subsidize the development of HVDC links between Ireland, Britain, the Netherlands, Germany, Denmark, and Sweden, as part of a wider €1.2 billion package supporting links to offshore wind farms and cross-border interconnectors throughout Europe. Meanwhile the recently founded Union of the Mediterranean has embraced a Mediterranean Solar Plan to import large amounts of concentrating solar power into Europe from North Africa and the Middle East.

Voltage Sourced Converters (VSC)

The development of higher rated insulated gate bipolar transistors (IGBT), gate turn-off thyristors (GTO) and integrated gate-commutated thyristors (IGCTs), has made smaller HVDC systems economical. These may be installed in existing AC grids for their role in stabilising power flow without the additional short-circuit current that would be produced by an additional AC transmission line. The manufacturer ABB calls this concept "HVDC Light", while Siemens calls a similar concept "HVDC PLUS" (Power Link Universal System) and Alstom call their product based upon this technology "HVDC MaxSine". They have extended the use of HVDC down to blocks as small as a few tens of megawatts and lines as short as a few score kilometres of overhead line. There are several different variants of Voltage-Sourced Converter (VSC) technology: most "HVDC Light" installations use pulse width modulation but the most recent installations, along with "HVDC PLUS" and "HVDC MaxSine", are based on multilevel switching. Multilevel switching has the advantage that it allows the harmonic filtering equipment to be reduced or eliminated altogether. At the moment, the AC harmonic filters of typical Line Commutated Converter stations cover nearly half of the area of the whole station.

Limitations

The amount of power that can be sent over a transmission line is limited. The origins of the limits vary depending on the length of the line. For a short line, the heating of conductors due to line losses sets a thermal limit. If too much current is drawn, conductors may sag too close to the ground, or conductors and equipment may be damaged by overheating. For intermediate-length lines on the order of 100 km (62 mi), the limit is set by the voltage drop in the line. For longer AC lines, system stability sets the limit to the power that can be transferred. Approximately, the power flowing over an AC line is proportional to the sine of the phase angle of the voltage at the receiving and transmitting ends. Since this angle varies depending on system loading and generation, it is undesirable for the angle to approach 90 degrees. Very approximately, the allowable product of line length and maximum load is proportional to the square of the system voltage. Series capacitors or phase-shifting transformers are used on long lines to improve stability. High-voltage direct current lines are restricted only by thermal and voltage drop limits, since the phase angle is not material to their operation.

Up to now, it has been almost impossible to foresee the temperature distribution along the cable route, so that the maximum applicable current load was usually set as a compromise between understanding of operation conditions and risk minimisation. The availability of industrial Distributed Temperature Sensing (DTS) systems that measure in real time temperatures all along the cable is a first step in monitoring the transmission system capacity. This monitoring solution is based on using passive optical fibres as temperature sensors, either integrated directly inside a high voltage cable or mounted externally on the cable insulation. A solution for overhead lines is also available. In this case the optical fibre is integrated into the core of a phase wire of overhead transmission lines (OPPC). The integrated Dynamic Cable Rating (DCR) or also called Real Time Thermal Rating (RTTR) solution enables not only to continuously monitor the temperature of a high voltage cable circuit in real time, but to safely utilise the existing network capacity to its maximum. Furthermore it provides the ability to the operator to predict the behaviour of the transmission system upon major changes made to its initial operating conditions.

Control

To ensure safe and predictable operation the components of the transmission system are controlled with generators, switches, circuit breakers and loads. The voltage, power, frequency, load factor, and reliability capabilities of the transmission system are designed to provide cost effective performance for the customers.

Load Balancing

The transmission system provides for base load and peak load capability, with safety and fault tolerance margins. The peak load times vary by region largely due to the industry mix. In very hot and very cold climates home air conditioning and heating loads have an effect on the overall load. They are typically highest in the late afternoon in the hottest part of the year and in mid-mornings and mid-evenings in the coldest part of the year. This makes the power requirements vary by the season and the time of day. Distribution system designs always take the base load and the peak load into consideration.

The transmission system usually does not have a large buffering capability to match the loads with the generation. Thus generation has to be kept matched to the load, to prevent overloading failures of the generation equipment.

Multiple sources and loads can be connected to the transmission system and they must be controlled to provide orderly transfer of power. In centralised power generation, only local control of generation is necessary, and it involves synchronisation of the generation units, to prevent large transients and overload conditions.

In distributed power generation the generators are geographically distributed and the process to bring them online and offline must be carefully controlled. The load control signals can either be sent on separate lines or on the power lines themselves. To load balance the voltage and frequency can be used as a signalling mechanism.

In voltage signalling, the variation of voltage is used to increase generation. The power added by any system increases as the line voltage decreases. This arrangement is stable in principle. Voltage based regulation is complex to use in mesh networks, since the individual components and setpoints would need to be reconfigured every time a new generator is added to the mesh.

In frequency signalling, the generating units match the frequency of the power transmission system. In droop speed control, if the frequency decreases, the power is increased. (The drop in line frequency is an indication that the increased load is causing the generators to slow down.) Wind turbines, v2g and other distributed storage and generation systems can be connected to the power grid, and interact with it to improve system operation.

Failure Protection

Under excess load conditions, the system can be designed to fail gracefully rather than all at once. Brownouts occur when the supply power drops below the demand. Blackouts occur when the supply fails completely.

Rolling blackouts (also called load shedding) are intentionally engineered electrical power outages, used to distribute insufficient power when the demand for electricity exceeds the supply.

Communications

Operators of long transmission lines require reliable communications for control of the power grid and, often, associated generation and distribution facilities. Fault-sensing protective relays at each end of the line must communicate to monitor the flow of power into and out of the protected line section so that faulted conductors or equipment can be quickly de-energised and the balance of the system restored. Protection of the transmission line from short circuits and other faults is usually so critical that common carrier telecommunications are insufficiently reliable, and in remote areas a common carrier may not be available. Communication systems associated with a transmission project may use:

- Microwaves
- Power line communication
- Optical fibres.

Rarely, and for short distances, a utility will use pilot-wires strung along the transmission line path. Leased circuits from common carriers are not preferred since availability is not under control of the electric power transmission organisation.

Transmission lines can also be used to carry data: this is called power-line carrier, or PLC. PLC signals can be easily received with a radio for the long wave range. Optical fibres can be included in the stranded conductors of a transmission line, in the overhead shield wires. These cables are known as optical ground wire (*OPGW*). Sometimes a standalone cable is used, all-dielectric self-supporting (*ADSS*) cable, attached to the transmission line cross arms. Some jurisdictions, such as Minnesota, prohibit energy transmission companies from selling surplus communication bandwidth or acting as a telecommunications common carrier. Where the regulatory structure permits, the utility can sell capacity in extra dark fibres to a common carrier, providing another revenue stream.

Electricity Market Reform

Some regulators regard electric transmission to be a natural monopoly and there are moves in many countries to separately regulate transmission. Spain was the first country to establish a regional transmission organisation. In that country transmission operations and market operations are controlled by separate companies. The transmission system operator is Red Eléctrica de España (REE) and the wholesale electricity market operator is Operador del Mercado Ibérico de Energía – Polo Español, S.A. (OMEL). Spain's transmission system is interconnected with those of France, Portugal, and Morocco.

In the United States and parts of Canada, electrical transmission companies operate independently of generation and distribution companies.

Cost of Electric Power Transmission

The cost of high voltage electricity transmission (as opposed to the costs of electricity distribution) is comparatively low, compared to all other costs arising in a consumer's electricity bill. In the UK transmission costs are about 0.2p/kWh compared to a delivered domestic price of around 10 p/kWh. Research evaluates the level of capital expenditure in the electric power T&D equipment market will be worth $128.9bn in 2011.

Merchant Transmission

Merchant transmission is an arrangement where a third party constructs and operates electric transmission lines through the franchise area of an unrelated utility. Operating merchant transmission projects in the United States include the Cross Sound Cable from Long Island, New York to New Haven, Connecticut, Neptune RTS Transmission Line from Sayreville, N.J., to Newbridge, N.Y, ITC Holdings, Inc. transmission system in the midwest, and Path 15 in California. Additional projects are in development or have been proposed throughout the United States.

There is only one unregulated or market interconnector in Australia: Basslink between Tasmania and Victoria. Two DC links originally implemented as market interconnectors Directlink and Murraylink have been converted to regulated interconnectors. NEMMCO. A major barrier to wider adoption of merchant transmission is the difficulty in identifying who benefits from the facility so that the beneficiaries will pay the toll. Also, it is difficult for a merchant transmission line to compete when the alternative transmission lines are subsidized by other utility businesses.

Health Concerns

Some large studies, including a large United States study, have failed to find any link between living near power lines and developing any sickness or diseases such as cancer. One old study from 1997 found that it did not matter how close you were to a power line or a sub-station, there was no increased risk of cancer or illness.

The mainstream scientific evidence suggests that low-power, low-frequency, electromagnetic radiation associated with household currents and high transmission power lines does not constitute a short or long term health hazard. Some studies, however, have found statistical correlations between various diseases and living or working near power lines. No adverse health effects have been substantiated for people not living close to powerlines.

There are established biological effects for acute *high* level exposure to magnetic fields well above 100 μT (1000 mG). In a residential setting, there is "limited evidence of carcinogenicity in humans and less than sufficient evidence for carcinogenicity in experimental animals", in particular, childhood leukaemia, *associated with* average exposure to residential power-frequency magnetic field above 0.3 (3 mG) to 0.4 μT (4 mG). These levels exceed average residential power-frequency magnetic fields in homes which are about 0.07 μT (0.7 mG) in Europe and 0.11 μT (1.1 mG) in North America.

Government Policy

Historically, local governments have exercised authority over the grid and have significant disincentives to take action that would benefit states other than their own. Localities with cheap electricity have a disincentive to making interstate commerce in electricity trading easier, since other regions will be able to compete for local energy and drive up rates. Some regulators in Maine for example do not wish to address congestion problems because the congestion serves to keep Maine rates low. Further, vocal local constituencies can block or slow permitting by pointing to visual impact, environmental, and perceived health concerns.

In the US, generation is growing 4 times faster than transmission, but big transmission upgrades require the coordination of multiple states, a multitude of interlocking permits, and cooperation between a significant portion of the 500 companies that own the grid. From a policy perspective, the control of the grid is balkanized, and even former energy secretary Bill Richardson refers to it as a *third world grid*. There have been efforts in the EU and US to confront the problem. The US national security interest in significantly growing transmission capacity drove passage of the 2005 energy act giving the Department of Energy the authority to approve transmission if states refuse to act. However, soon after using its power to designate two

National Interest Electric Transmission Corridors, 14 senators signed a letter stating the DOE was being too aggressive.

Traction Power Network

A traction network or traction power network is an electricity grid for the supply of electrified rail networks. The installation of a separate traction network generally is only done if the railway in question uses alternating current (AC) with a frequency lower than that of the national grid, such as in Germany, Austria and Switzerland.

Alternatively, the three-phase alternating current of the power grid can be converted in substations by rotary transformers or static inverters into the voltage and type of current required by the trains. For railways which run on direct current (DC), this method is always used, as well as for railways which run on single-phase AC of decreased frequency, as in Mecklenburg-Western Pomerania, Saxony-Anhalt, Norway and Sweden. In these areas there are no traction current networks.

History

Separate power for traction apart from industrial power always has historic roots. There is no reason today to apply different frequencies or current types than for transmission and for industrial usage. However, the advantage with DC traction was the easier transmission with single copper wires to the feeder points. The advantage with AC traction is the easier transmission over long distances to the feeder points. Beyond these parameters and securing former investment, no evidence exists to stay with different current schemes in networks.

Applications

Dedicated traction current lines are used when railways are supplied with low-frequency alternating current (AC). The traction current supply line is connected to substations along the line of the railway and is usually run separately from the overhead catenary wire from which the locomotives are fed.

In countries in which the electric trains run with direct current or with single-phase alternating current with the frequency of the general power grid, the required conversion of the current is performed in the substations, so again no traction current lines are required.

Traction current supply lines are not usually laid parallel to the railway line, in order to allow a shorter line length and to avoid unnecessary influences to the electrical system near the railway line;

this also is applied to the current supply of some rapid-transit railways operating with alternating current in Germany.

It is also possible to lay out the traction current supply on special cross beams right on the overhead wire pylons above the catenary wire. Because the overhead line pylons have a smaller cross section than traction current supply masts, the cross beams cannot be too wide, so the standard arrangement of four conductor cables in one level cannot be used. In this case, a two-level arrangement is used, or with two electric circuits for double-railed lines the overhead line pylons for both directions are equipped with cross beams for their own traction current system of two conductor cables each. In densely populated areas, there are pylons which carry circuits for both traction current and for three-phase alternating current for general power. Such lines are found where right of ways are rare. In particular the parallel route of 110 kV and 220 kV three-phase AC is common. The use of 380 kV power lines on the same pylon requires 220 kV insulators for the traction current line, because in case the 380 kV line fails, voltage spikes can occur along the traction current line, which the 110 kV insulators cannot handle. As a rule, traction current lines use single conductors, however for the supply of railways with high traffic and in particular for the supply of high speed railway lines, two bundle conductors are used.

Around the World

Austria

The Mariazeller railway in Lower Austria operates on single phase AC at a 25 Hz utility frequency. The railway has its own traction current lines with an operating voltage of 27 kV. These lines are mounted on the pylons of the overhead wire over the catenary wire.

Germany

In Germany, single conductors are usually used for traction current lines but, for the ICE train, two bundle conductors are used. The traction current supply lines from the nuclear power station Neckarwestheim to the traction current switching station at Neckarwestheim and from there to the central substation in Stuttgart, Zazenhausen are implemented as a four-bundle conductor circuit.

Scandinavia

In Sweden, Norway and some areas of the former German Democratic Republic, three phase AC-current is converted into single phase AC current with a frequency of 16.7 cycles per second at the substations. Unlike in Germany, there are no dedicated power plants for railway electricity. All power comes from general electricity suppliers. Although in this region there is, in principle, no requirement for traction power lines, there is a 132 kV-single AC power grid for railway power supply in Central Sweden. In Norway, there is a small 55 kV single phase AC network for power supply of trains in the South, fed by Hakavik Power Station.

A further power station, at Kjofossen feeds single phase AC directly in the overhead wire. In Denmark and Finland, 50 Hz is used for the main lines (if electrified) and the electricity comes from general suppliers. As such, much simpler equipment than in Sweden and Norway is needed for conversion.

South Africa

In the Republic of South Africa there are extensive AC and DC traction schemes, including 50 kV and 25 kV AC single phase systems. Electrification in Natal was stimulated by the takeover of the South African Railways' system by the Electricity Supply Commission (now Eskom) based on the Colenso power station.

United Kingdom

In the United Kingdom, the Network Rail 750 V DC electrification system in the southeast of England is supplied with power from an extensive 33 kV power distribution network.

Areas with Traction Power Networks

- United Kingdom
- Germany (except Mecklenburg-Western Pomerania and Saxony-Anhalt), total length 7959 km
- Switzerland
- Austria (separate traction power network for the Mariazeller Bahn)
- Central Sweden
- Southern Norway east of Oslo
- USA (in New York and Washington DC area for railway lines running with single phase 25 Hz AC)
- South Africa
- Melbourne, Australia (dedicated high voltage transmission wires between former Newport A Power Station and traction

substations. However these lines are operated with three phase AC).

Superconducting Cables

High-temperature superconductors promise to revolutionise power distribution by providing lossless transmission of electrical power. The development of superconductors with transition temperatures higher than the boiling point of liquid nitrogen has made the concept of superconducting power lines commercially feasible, at least for high-load applications. It has been estimated that the waste would be halved using this method, since the necessary refrigeration equipment would consume about half the power saved by the elimination of the majority of resistive losses. Some companies such as Consolidated Edison and American Superconductor have already begun commercial production of such systems. In one hypothetical future system called a SuperGrid, the cost of cooling would be eliminated by coupling the transmission line with a liquid hydrogen pipeline. Superconducting cables are particularly suited to high load density areas such as the business district of large cities, where purchase of an easement for cables would be very costly.

Single wire Earth Return

Single-wire earth return (SWER) or single wire ground return is a single-wire transmission line for supplying single-phase electrical power for an electrical grid to remote areas at low cost. It is principally used for rural electrification, but also finds use for larger isolated loads such as water pumps, and light rail. Single wire earth return is also used for HVDC over submarine power cables.

Wireless Power Transmission

Both Nikola Tesla and Hidetsugu Yagi attempted to devise systems for large scale wireless power transmission, with no commercial success.

In November 2009, LaserMotive won the NASA 2009 Power Beaming Challenge by powering a cable climber 1 km vertically using a ground-based laser transmitter. The system produced up to 1 kW of power at the receiver end. In August 2010, NASA contracted with private companies to pursue the design of laser power beaming systems to power low earth orbit satellites and to launch rockets using laser power beams.

Wireless power transmission has been studied for transmission of power from solar power satellites to the earth. A high power array of microwave or laser transmitters would beam power to a rectenna.

Major engineering and economic challenges face any solar power satellite project.

Security of Control Systems

The Federal government of the United States admits that the power grid is susceptible to cyber-warfare. The United States Department of Homeland Security works with industry to identify vulnerabilities and to help industry enhance the security of control system networks, the federal government is also working to ensure that security is built in as the U.S. develops the next generation of 'smart grid' networks.

Records

- Highest capacity system: 6.3 GW HVDC Itaipu (Brazil) (±600 kV DC)
- Highest transmission voltage (AC): 1.15 MV on Powerline Ekibastuz-Kokshetau (Kazakhstan)
- Largest double-circuit transmission, Kita-Iwaki Powerline.
- Highest pylons: Yangtze River Crossing (height: 345 m/1,132 ft)
- Longest power line: Inga-Shaba (length: 1,700 kilometres / 1,056 miles)
- Longest span of power line: 5,376 m (17,638 ft) at Ameralik Span
- Longest submarine cables:
    - NorNed, North Sea – (length of submarine cable: 580 kilometres / 360 miles)
    - Basslink, Bass Strait – (length of submarine cable: 290 kilometres / 180 miles, total length: 370.1 kilometres / 230 miles)
    - Baltic-Cable, Baltic Sea – (length of submarine cable: 238 kilometres / 148 miles, HVDC length: 250 kilometres / 155 miles, total length: 262 kilometres / 163 miles)
- Longest underground cables:
    - Murraylink, Riverland/Sunraysia – (length of underground cable: 180 kilometres / 112 miles).

Chapter 2

Power System Automation

Power system automation is the act of automatically controlling the power system via instrumentation and control devices. Substation automation refers to using data from Intelligent electronic devices (IED), control and automation capabilities within the substation, and control commands from remote users to control power system devices.

Since full substation automation relies on substation integration, the terms are often used interchangeably. Power system automation includes processes associated with generation and delivery of power. Monitoring and control of power delivery systems in the substation and on the pole top reduce the occurrence of outages and shorten the duration of outages that do occur. The IEDs, communications protocols, and communications methods, work together as a system to perform power system automation. The term "power system" describes the collection of devices that make up the physical systems that generate, transmit, and distribute power. The term "instrumentation and control (I&C) system" refers to the collection of devices that monitor, control, and protect the power system.

Automation Tasks

Power system automation is composed of several tasks.

Data Acquisition : Data acquisition refers to acquiring, or collecting, data. This data is collected in the form of measured analog current or voltage values or the open or closed status of contact points. Acquired data can be used locally within the device collecting it, sent to another device in a substation, or sent from the substation to one or several databases for use by operators, engineers, planners, and administration.

Supervision : Computer processes and personnel supervise, or monitor, the conditions and status of the power system using this acquired data. Operators and engineers monitor the information remotely on computer displays and graphical wall displays or locally, at the device, on front-panel displays and laptop computers.

Control : Control refers to sending command messages to a device to operate the I&C and power system devices. Traditional supervisory control and data acquisition (SCADA) systems rely on operators to supervise the system and initiate commands from an operator console on the master computer. Field personnel can also control devices using front-panel push buttons or a laptop computer.

In addition, another task is power system integration, which is the act of communicating data to, from, or among IEDs in the I&C system and remote users. Substation integration refers to combining data from the IED's local to a substation so that there is a single point of contact in the substation for all of the I&C data.

The power system automation processes rely on data acquisition; Power system supervision and power system control all working together in a coordinated automatic fashion. The commands are generated automatically and then transmitted in the same fashion as operator initiated commands.

Hardware Structure of the Power System Automation

Data Acquisition

The instruments transformers with protective relays are used to sense the power system voltage and current. They are physically connected to power system apparatus and convert the actual power system signals. The transducers convert the analog output of an instrument transformer from one magnitude to another or from one value type to another, such as from an ac current to dc voltage. Also the input data is taken from the auxiliary contacts of switch gears and power system control equipments.

Main processing instrumentation and control (I&C) device The I&C devices built using microprocessors are commonly referred to as intelligent electronic devices (IEDs). Microprocessors are single chip computers that allow the devices into which they are built to process data, accept commands, and communicate information like a computer. Automatic processes can be run in the IEDs. Some IEDs used in power system automation are:

Remote Terminal Unit, RTU: A remote terminal unit is an IED that can be installed in a remote location, and acts as a termination point for field contacts. A dedicated pair of copper conductors is used to sense every contact and transducer value. These conductors originate at the power system device, are installed in trenches or overhead cable trays, and are then terminated on panels within the RTU. The RTU can transfer collected data to other devices and receive data and control commands from other devices. User programmable RTUs are referred to as "smart RTUs."

Meter : A meter is an IED that is used to create accurate measurements of power system current, voltage, and power values. Metering values such as demand and peak are saved within the meter to create historical information about the activity of the power system.

Digital Fault Recorder : A digital fault recorder (DFR) is an IED that records information about power system disturbances. It is capable of storing data in a digital format when triggered by conditions detected on the power system. Harmonics, frequency, and voltage are examples of data captured by DFRs.

Programmable Logic Controller (PLC) : A Programmable Logic Controller (PLC) can be programmed to perform logical control. As with the RTU, a dedicated pair of copper conductors for each contact and transducer value is terminated on panels within the PLC. It is like a work-horse which work upon the command given by their master.

Protective Relay : A protective relay is an IED designed to sense power system disturbances and automatically perform control actions on the I&C system and the power system to protect personnel and equipment. The relay has local termination so that the copper conductors for each contact do not have to be routed to a central termination panel associated with RTU

Controlling (Output) Devices

Load Tap Changer (LTC) : Load tap changers are devices used to change the tap position on transformers. These devices work automatically or can be controlled via another local IED or from a remote operator or process.

Recloser Controller : Recloser controllers remotely control the operation of automated reclosers and switches. These devices monitor and store power system conditions and determine when to perform control actions. They also accept commands from a remote operator or process.

Communications Devices

Communications Processor : A communications processor is a substation controller that incorporates the functions of many other

I&C devices into one IED. It has many communications ports to support multiple simultaneous communications links. The communications processor performs data acquisition and control of the other substation IEDs and also concentrates the data it acquires for transmission to one or many masters inside and outside the substation.

Applications

Overcurrent Protection

All lines and all electrical equipment must be protected against prolonged overcurrent. If the cause of the overcurrent is nearby then automatically that current is interrupted immediately. But if the cause of the overcurrent is outside the local area then a backup provision automatically disconnects all affected circuits after a suitable time delay.

Note that disconnection can, unfortunately, have a cascade effect, leading to overcurrent in other circuits that then also must therefore disconnect automatically.

Also note that generators that suddenly have lost their load because of such a protection operation will have to shut down automatically immediately, and it may take many hours to restore a proper balance between demand and supply in the system, partly because there must be proper synchronisation before any two parts of the system can be reconnected. Reclosing operations of circuit breakers usually are attempted automatically, and often are successful during thunderstorms, for example.

Supervisory Control and Data Acquisition

It is the system which transmits and receives logic or data from events of controls, metering, measuring, safety and monitoring of process devices such as Electrical equipment, Instrumentation devices, telecommunication on industrial applications. Some power stations can control other stations via communication links over long distances.

Remote switching, telemetering of home grids (showing voltage, current, power, direction, consumption in kWh, etc.), even automatic synchronisation is used in some power systems.

AC Power

Power in an electric circuit is the rate of flow of energy past a given point of the circuit. In alternating current circuits, energy storage elements such as inductance and capacitance may result in periodic reversals of the direction of energy flow. The portion of power that, averaged over a complete cycle of the AC waveform, results in net transfer of energy in one direction is known as real power.

The portion of power due to stored energy, which returns to the source in each cycle, is known as reactive power.

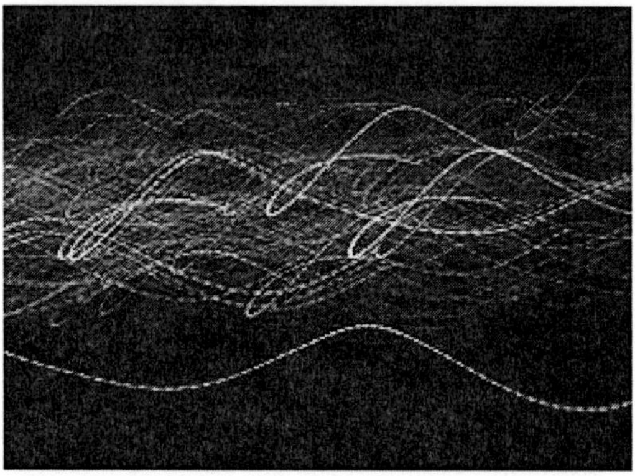

Figure : Usually hidden from the unaided eye, the blinking of (non-incandescent) lighting powered by AC mains is revealed in this motion-blurred long exposure of city lights. Light is emitted twice each cycle.

Real, reactive, and apparent powers

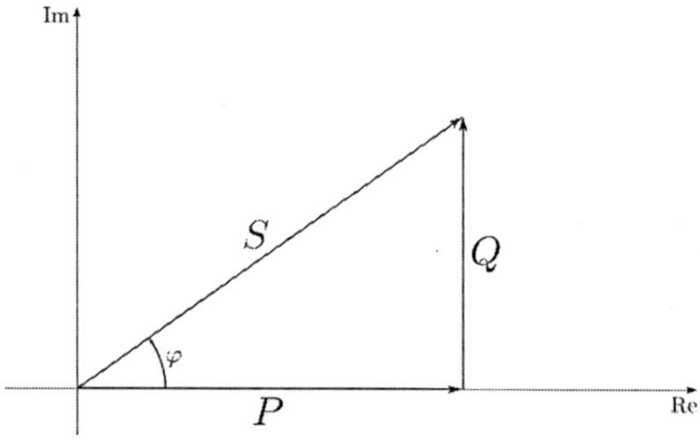

Figure : The complex power is the vector sum of real and reactive power. The apparent power is the magnitude of the complex power.

Real power (P)

Reactive power (Q)

Complex power (S)

Apparent Power (|S|)

Phase of Current (φ)

In a simple alternating current (AC) circuit consisting of a source and a linear load, both the current and voltage are sinusoidal. If the load is purely resistive, the two quantities reverse their polarity at the same time. At every instant the product of voltage and current is positive, indicating that the direction of energy flow does not reverse. In this case, only real power is transferred.

If the loads are purely reactive, then the voltage and current are 90 degrees out of phase. For half of each cycle, the product of voltage and current is positive, but on the other half of the cycle, the product is negative, indicating that on average, exactly as much energy flows toward the load as flows back. There is no net energy flow over one cycle. In this case, only reactive energy flows—there is no net transfer of energy to the load.

Practical loads have resistance, inductance, and capacitance, so both real and reactive power will flow to real loads. Power engineers measure apparent power as the magnitude of the vector sum of real and reactive power. Apparent power is the product of the root-mean-square of voltage and current.

Engineers care about apparent power, because even though the current associated with reactive power does no work at the load, it heats the wires, wasting energy. Conductors, transformers and generators must be sized to carry the total current, not just the current that does useful work.

Another consequence is that adding the apparent power for two loads will not accurately give the total apparent power unless they have the same displacement between current and voltage (the same power factor).

Conventionally, capacitors are considered to generate reactive power and inductors to consume it. If a capacitor and an inductor are placed in parallel, then the currents flowing through the inductor and the capacitor tend to cancel rather than add. This is the fundamental mechanism for controlling the power factor in electric power transmission; capacitors (or inductors) are inserted in a circuit to partially cancel reactive power 'consumed' by the load.

Engineers use the following terms to describe energy flow in a system (and assign each of them a different unit to differentiate between them):

- Real power (P) or active power: watt [W]
- Reactive power (Q): volt-ampere reactive [var]
- Complex power (S): volt-ampere [VA]
- Apparent Power ($|S|$), that is, the absolute value of complex power S: volt-ampere [VA]

- Phase of Voltage Relative to Current (φ), the angle of difference (in degrees) between voltage and current; Current lagging Voltage (Quadrant I Vector), Current leading voltage (Quadrant IV Vector)

In the diagram, P is the real power, Q is the reactive power (in this case positive), S is the complex power and the length of S is the apparent power. Reactive power does not transfer energy, so it is represented as the imaginary axis of the vector diagram. Real power moves energy, so it is the real axis. The unit for all forms of power is the watt (symbol: W), but this unit is generally reserved for real power. Apparent power is conventionally expressed in volt-amperes (VA) since it is the product of rms voltage and rms current. The unit for reactive power is expressed as var, which stands for volt-amperes reactive. Since reactive power transfers no net energy to the load, it is sometimes called "wattless" power. It does, however, serve an important function in electrical grids and its lack has been cited as a significant factor in the Northeast Blackout of 2003. Understanding the relationship between these three quantities lies at the heart of understanding power engineering. The mathematical relationship among them can be represented by vectors or expressed using complex numbers, $S = P + jQ$ (where j is the imaginary unit).

Power Factor

The power factor of an AC electric power system is defined as the ratio of the real power flowing to the load to the apparent power in the circuit, and is a dimensionless number between 0 and 1. Real power is the capacity of the circuit for performing work in a particular time. Apparent power is the product of the current and voltage of the circuit. Due to energy stored in the load and returned to the source, or due to a non-linear load that distorts the wave shape of the current drawn from the source, the apparent power will be greater than the real power. In an electric power system, a load with a low power factor draws more current than a load with a high power factor for the same amount of useful power transferred. Linear loads with low power factor (such as induction motors) can be corrected with a passive network of capacitors or inductors. Non-linear loads, such as rectifiers, distort the current drawn from the system. In such cases, active or passive power factor correction may be used to counteract the distortion and raise the power factor. The devices for correction of the power factor may be at a central substation, spread out over a distribution system, or built into power-consuming equipment.

Linear Circuits

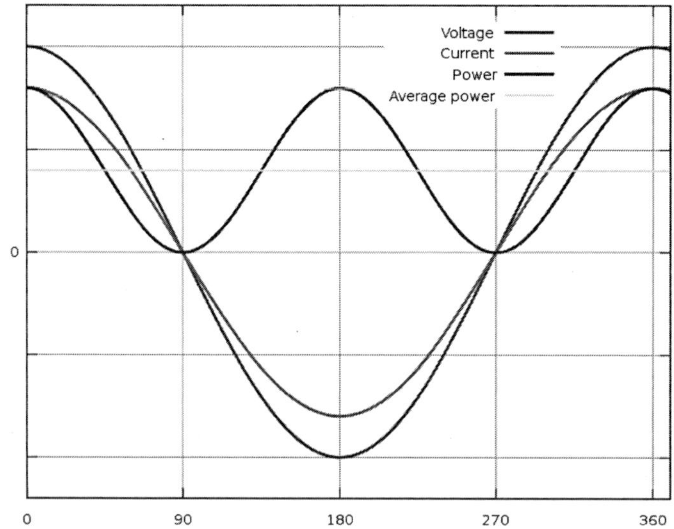

Figure : Instantaneous and average power calculated from AC voltage and current with a unity power factor (*φ = 0, cos φ = 1*). Since the blue line is above the axis, all power is real power consumed by the load.

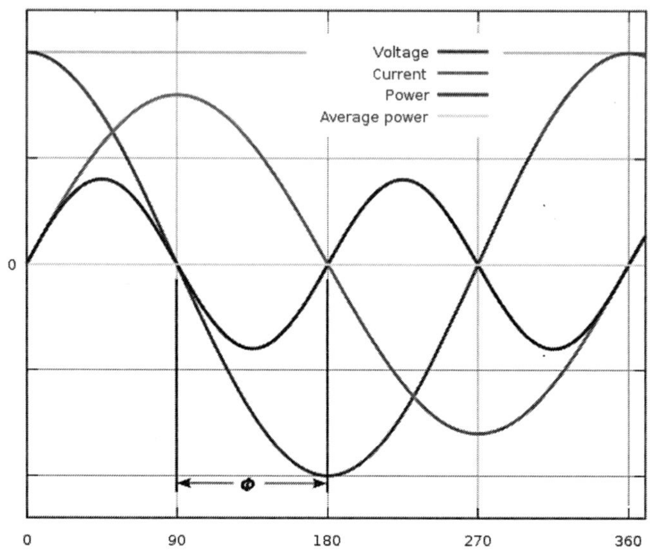

Figure : Instantaneous and average power calculated from AC voltage and current with a zero power factor (*φ – 90, cos φ = 0*). The blue line shows all the power is stored temporarily in the load during the first quarter cycle and returned to the grid during the second quarter cycle, so no real power is consumed.

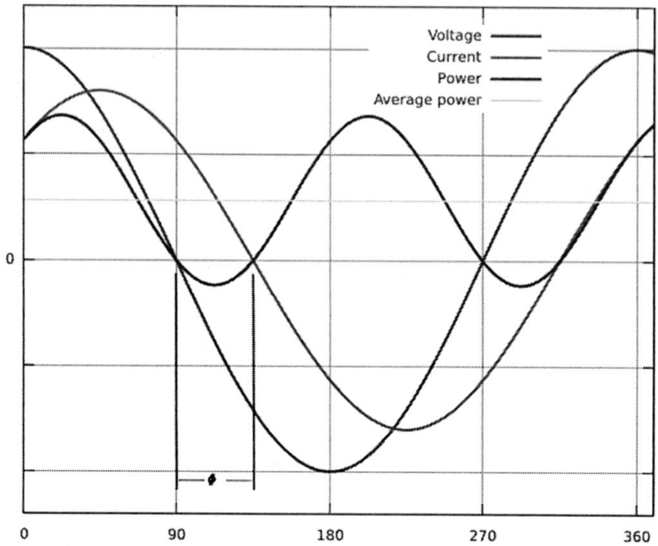

Figure : Instantaneous and average power calculated from AC voltage and current with a lagging power factor ($\varphi = 45, cos\varphi = 0.71$). The blue line shows some of the power is returned to the grid during the part of the cycle labelled φ.

In a purely resistive AC circuit, voltage and current waveforms are in step (or in phase), changing polarity at the same instant in each cycle. All the power entering the load is consumed. Where reactive loads are present, such as with capacitors or inductors, energy storage in the loads result in a time difference between the current and voltage waveforms.

During each cycle of the AC voltage, extra energy, in addition to any energy consumed in the load, is temporarily stored in the load in electric or magnetic fields, and then returned to the power grid a fraction of a second later in the cycle. The "ebb and flow" of this nonproductive power increases the current in the line.

Thus, a circuit with a low power factor will use higher currents to transfer a given quantity of real power than a circuit with a high power factor. A linear load does not change the shape of the waveform of the current, but may change the relative timing (phase) between voltage and current.

Circuits containing purely resistive heating elements (filament lamps, cooking stoves, etc.) have a power factor of 1.0. Circuits containing inductive or capacitive elements (electric motors, solenoid valves, lamp ballasts, and others) often have a power factor below 1.0.

Definition and Calculation

AC power flow has the three components: real power (also known as active power) (P), measured in watts (W); apparent power (S), measured in volt-amperes (VA); and reactive power (Q), measured in reactive volt-amperes (var).

The power factor is defined as:

$$\frac{P}{S}$$

In the case of a perfectly sinusoidal waveform, P, Q and S can be expressed as vectors that form a vector triangle such that:

$$S^2 = P^2 + Q^2.$$

If φ is the phase angle between the current and voltage, then the power factor is equal to the cosine of the angle, $|\cos \varphi|$, and:

$$|P| = |S|\,|\cos\varphi|.$$

Since the units are consistent, the power factor is by definition a dimensionless number between 0 and 1. When power factor is equal to 0, the energy flow is entirely reactive, and stored energy in the load returns to the source on each cycle. When the power factor is 1, all the energy supplied by the source is consumed by the load. Power factors are usually stated as "leading" or "lagging" to show the sign of the phase angle.

If a purely resistive load is connected to a power supply, current and voltage will change polarity in step, the power factor will be unity (1), and the electrical energy flows in a single direction across the network in each cycle. Inductive loads such as transformers and motors (any type of wound coil) consume reactive power with current waveform lagging the voltage. Capacitive loads such as capacitor banks or buried cable generate reactive power with current phase leading the voltage. Both types of loads will absorb energy during part of the AC cycle, which is stored in the device's magnetic or electric field, only to return this energy back to the source during the rest of the cycle.

For example, to get 1 kW of real power, if the power factor is unity, 1 kVA of apparent power needs to be transferred (1 kW ÷ 1 = 1 kVA). At low values of power factor, more apparent power needs to be transferred to get the same real power. To get 1 kW of real power at 0.2 power factor, 5 kVA of apparent power needs to be transferred

(1 kW ÷ 0.2 = 5 kVA). This apparent power must be produced and transmitted to the load in the conventional fashion, and is subject to the usual distributed losses in the production and transmission processes.

Electrical loads consuming alternating current power consume both real power and reactive power. The vector sum of real and reactive power is the apparent power. The presence of reactive power causes the real power to be less than the apparent power, and so, the electric load has a power factor of less than 1.

Power Factor Correction of Linear Loads

A high power factor is generally desirable in a transmission system to reduce transmission losses and improve voltage regulation at the load. It is often desirable to adjust the power factor of a system to near 1.0. When reactive elements supply or absorb reactive power near the load, the apparent power is reduced. Power factor correction may be applied by an electrical power transmission utility to improve the stability and efficiency of the transmission network. Individual electrical customers who are charged by their utility for low power factor may install correction equipment to reduce those costs.

Power factor correction brings the power factor of an AC power circuit closer to 1 by supplying reactive power of opposite sign, adding capacitors or inductors that act to cancel the inductive or capacitive effects of the load, respectively. For example, the inductive effect of motor loads may be offset by locally connected capacitors. If a load had a capacitive value, inductors (also known as *reactors* in this context) are connected to correct the power factor. In the electricity industry, inductors are said to *consume* reactive power and capacitors are said to *supply* it, even though the reactive power is actually just moving back and forth on each AC cycle.

The reactive elements can create voltage fluctuations and harmonic noise when switched on or off. They will supply or sink reactive power regardless of whether there is a corresponding load operating nearby, increasing the system's no-load losses. In the worst case, reactive elements can interact with the system and with each other to create resonant conditions, resulting in system instability and severe overvoltage fluctuations. As such, reactive elements cannot simply be applied without engineering analysis.

An automatic power factor correction unit consists of a number of capacitors that are switched by means of contactors. These contactors are controlled by a regulator that measures power factor in an electrical network. Depending on the load and power factor of the network, the power factor controller will switch the necessary blocks of capacitors in steps to make sure the power factor stays above a selected value.

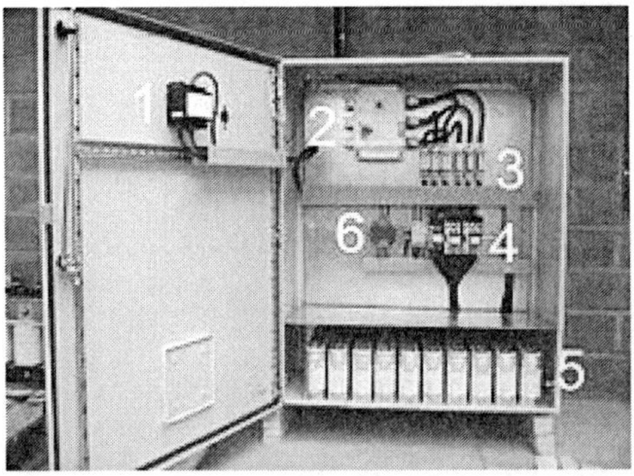

Figure : Reactive Power Control Relay; 2. Network connection points; 3. Slow-blow Fuses; 4. Inrush Limiting Contactors; 5. Capacitors (single-phase or three-phase units, delta-connection); 6. Transformer Suitable voltage transformation to suit control power (contactors, ventilation,...)

Instead of using a set of switched capacitors, an unloaded synchronous motor can supply reactive power. The reactive power drawn by the synchronous motor is a function of its field excitation. This is referred to as a *synchronous condenser*.

It is started and connected to the electrical network. It operates at a leading power factor and puts vars onto the network as required to support a system's voltage or to maintain the system power factor at a specified level.

The condenser's installation and operation are identical to large electric motors. Its principal advantage is the ease with which the amount of correction can be adjusted; it behaves like an electrically variable capacitor. Unlike capacitors, the amount of reactive power supplied is proportional to voltage, not the square of voltage; this improves voltage stability on large networks. Synchronous condensors are often used in connection with high-voltage direct-current transmission projects or in large industrial plants such as steel mills.

For power factor correction of high-voltage power systems or large, fluctuating industrial loads, power electronic devices such as the Static VAR compensator or STATCOM are increasingly used. These systems are able to compensate sudden changes of power factor much more rapidly than contactor-switched capacitor banks, and being solid-state require less maintenance than synchronous condensers.

Non-linear loads

A non-linear load on a power system is typically a rectifier (such as used in a power supply), or some kind of arc discharge device such as a fluorescent lamp, electric welding machine, or arc furnace. Because current in these systems is interrupted by a switching action, the current contains frequency components that are multiples of the power system frequency. Distortion power factor is a measure of how much the harmonic distortion of a load current decreases the average power transferred to the load.

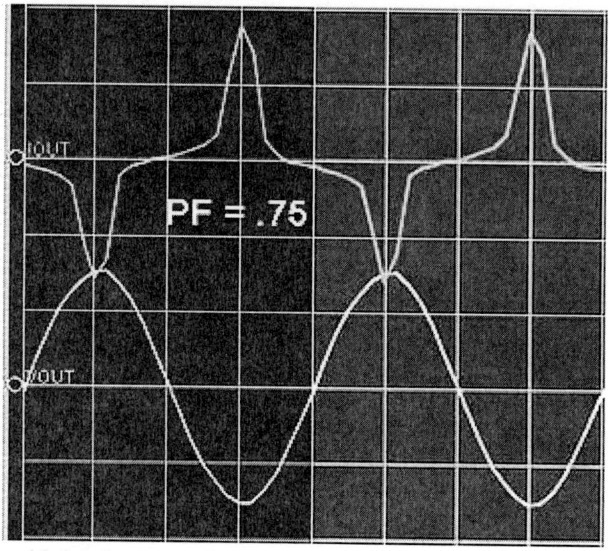

Figure : Sinusoidal voltage and non-sinusoidal current give a distortion power factor of 0.75 for this computer power supply load.

Non-sinusoidal Components

Non-linear loads change the shape of the current waveform from a sine wave to some other form. Non-linear loads create harmonic currents in addition to the original (fundamental frequency) AC current. Filters consisting of linear capacitors and inductors can prevent harmonic currents from entering the supplying system.

In linear circuits having only sinusoidal currents and voltages of one frequency, the power factor arises only from the difference in phase between the current and voltage. This is "displacement power factor". The concept can be generalised to a total, distortion, or true power factor where the apparent power includes all harmonic components. This is of importance in practical power systems that contain non-linear loads such as rectifiers, some forms of electric lighting, electric arc furnaces, welding equipment, switched-mode power supplies and other devices.

A typical multimetre will give incorrect results when attempting to measure the AC current drawn by a non-sinusoidal load; the instruments sense the average value of a rectified waveform. The average response is then calibrated to the effective, RMS value. An RMS sensing multimeter must be used to measure the actual RMS currents and voltages (and therefore apparent power). To measure the real power or reactive power, a watt meter designed to work properly with non-sinusoidal currents must be used.

Distortion Power Factor

The *distortion power factor* describes how the harmonic distortion of a load current decreases the average power transferred to the load.

$$\text{distortion power factor} = \frac{1}{\sqrt{1+\text{THD}_i^2}} = \frac{I_{1,\,\text{rms}}}{I_{\text{rms}}}$$

THD_i is the total harmonic distortion of the load current. This definition assumes that the voltage stays undistorted (sinusoidal, without harmonics). This simplification is often a good approximation in practice. $I_{1,\text{rms}}$ is the fundamental component of the current and I_{rms} is the total current-both are root mean square-values.

The result when multiplied with the displacement power factor (DPF) is the overall, true power factor or just power factor (PF):

$$\text{PF} = \text{DPF}\frac{I_{1,\,\text{rms}}}{I_{\text{rms}}}$$

Switched-mode Power Supplies

A particularly important class of non-linear loads is the millions of personal computers that typically incorporate switched-mode power supplies (SMPS) with rated output power ranging from a few watts to more than 1 kW. Historically, these very-low-cost power supplies incorporated a simple full-wave rectifier that conducted only when the mains instantaneous voltage exceeded the voltage on the input capacitors. This leads to very high ratios of peak-to-average input current, which also lead to a low distortion power factor and potentially serious phase and neutral loading concerns.

A typical switched-mode power supply first makes a DC bus, using a bridge rectifier or similar circuit. The output voltage is then derived from this DC bus. The problem with this is that the rectifier is a non-linear device, so the input current is highly non-linear. That means that the input current has energy at harmonics of the frequency of the voltage.

This presents a particular problem for the power companies, because they cannot compensate for the harmonic current by adding simple capacitors or inductors, as they could for the reactive power drawn by a linear load. Many jurisdictions are beginning to legally require power factor correction for all power supplies above a certain power level.

Regulatory agencies such as the EU have set harmonic limits as a method of improving power factor. Declining component cost has hastened implementation of two different methods. To comply with current EU standard EN61000-3-2, all switched-mode power supplies with output power more than 75 W must include passive PFC, at least. 80 PLUS power supply certification requires a power factor of 0.9 or more.

Power Factor Correction in Non-linear Loads

Passive PFC

The simplest way to control the harmonic current is to use a filter: it is possible to design a filter that passes current only at line frequency (e.g. 50 or 60 Hz). This filter reduces the harmonic current, which means that the non-linear device now looks like a linear load. At this point the power factor can be brought to near unity, using capacitors or inductors as required. This filter requires large-value high-current inductors, however, which are bulky and expensive.

A passive PFC requires an inductor larger than the inductor in an active PFC, but costs less. This is a simple way of correcting the nonlinearity of a load by using capacitor banks. It is not as effective as active PFC. Passive PFCs are typically more power efficient than active PFCs. *Efficiency* is not to be confused with the PFC, though many computer hardware reviews conflate them. A passive PFC on a switching computer PSU has a typical power efficiency of around 96%, while an active PFC has a typical efficiency of about 94%.

Active PFC

An "active power factor corrector" (active PFC) is a power electronic system that controls the amount of power drawn by a load in order to obtain a power factor as close as possible to unity. In most applications, the active PFC controls the input current of the load so that the current waveform is proportional to the mains voltage waveform (a sine wave). The purpose of making the power factor as close to unity (1) as possible is to make the load circuitry that is power factor corrected appear purely resistive (apparent power equal to real power). In this case, the voltage and current are in phase and the reactive power consumption is zero. This enables the most efficient delivery of electrical power from the power company to the consumer.

Some types of active PFC are:

- Boost
- Buck
- Buck-boost

Active power factor correctors can be single-stage or multi-stage.

In the case of a switched-mode power supply, a boost converter is inserted between the bridge rectifier and the main input capacitors. The boost converter attempts to maintain a constant DC bus voltage on its output while drawing a current that is always in phase with and at the same frequency as the line voltage. Another switch mode converter inside the power supply produces the desired output voltage from the DC bus. This approach requires additional semiconductor switches and control electronics, but permits cheaper and smaller passive components. It is frequently used in practice. For example, SMPS with passive PFC can achieve power factor of about 0.7–0.75, SMPS with active PFC, up to 0.99 power factor, while a SMPS without any power factor correction has a power factor of only about 0.55–0.65.

Due to their very wide input voltage range, many power supplies with active PFC can automatically adjust to operate on AC power from about 100 V (Japan) to 230 V (Europe). That feature is particularly welcome in power supplies for laptops.

Importance of Power Factor in Distribution Systems

The significance of power factor lies in the fact that utility companies supply customers with volt-amperes, but bill them for watts. Power factors below 1.0 require a utility to generate more than the minimum volt-amperes necessary to supply the real power (watts). This increases generation and transmission costs. For example, if the load power factor were as low as 0.7, the apparent power would be 1.4 times the real power used by the load. Line current in the circuit would also be 1.4 times the current required at 1.0 power factor, so the losses in the circuit would be doubled (since they are proportional to the square of the current). Alternatively all components of the system such as generators, conductors, transformers, and switchgear would be increased in size (and cost) to carry the extra current.

Utilities typically charge additional costs to customers who have a power factor below some limit, which is typically 0.9 to 0.95. Engineers are often interested in the power factor of a load as one of the factors that affect the efficiency of power transmission.

With the rising cost of energy and concerns over the efficient delivery of power, active PFC has become more common in consumer electronics. Current Energy Star guidelines for computers (ENERGY STAR Program Requirements for Computers Version 5.0) call for a power factor of ≥ 0.9 at 100% of rated output in the PC's power supply. According to a white paper authored by Intel and the U.S. Environmental Protection Agencyý, PCs with internal power supplies will require the use of active power factor correction to meet the ENERGY STAR 5.0 Program Requirements for Computers. In Europe, IEC 555-2 requires power factor correction be incorporated into consumer products.

Measuring Power Factor

Power factor in a single-phase circuit (or balanced three-phase circuit) can be measured with the wattmeter-ammeter-voltmeter method, where the power in watts is divided by the product of measured voltage and current. The power factor of a balanced polyphase circuit is the same as that of any phase. The power factor of an unbalanced polyphase circuit is not uniquely defined. A direct reading power factor meter can be made with a moving coil meter of the electrodynamic type, carrying two perpendicular coils on the moving part of the instrument. The field of the instrument is energised by the circuit current flow. The two moving coils, A and B, are connected in parallel with the circuit load. One coil, A, will be connected through a resistor and the second coil, B, through an inductor, so that the current in coil B is delayed with respect to current in A. At unity power factor, the current in A is in phase with the circuit current, and coil A provides maximum torque, driving the instrument pointer toward the 1.0 mark on the scale. At zero power factor, the current in coil B is in phase with circuit current, and coil B provides torque to drive the pointer towards 0. At intermediate values of power factor, the torques provided by the two coils add and the pointer takes up intermediate positions.

Another electromechanical instrument is the polarized-vane type. In this instrument a stationary field coil produces a rotating magnetic field, just like a polyphase motor. The field coils are connected either directly to polyphase voltage sources or to a phase-shifting reactor if a single-phase application. A second stationary field coil, perpendicular to the voltage coils, carries a current proportional to current in one phase of the circuit. The moving system of the instrument consists of two vanes that are magnetised by the current coil. In operation the moving vanes take up a physical angle equivalent to the electrical angle between the voltage source and the current source. This type of instrument can be made to register for currents in both directions, giving a four-quadrant display of power factor or phase angle.

Digital instruments can be made that either directly measure the time lag between voltage and current waveforms and so calculate the power factor, or by measuring both true and apparent power in the circuit and calculating the quotient. The first method is only accurate if voltage and current are sinusoidal; loads such as rectifiers distort the waveforms from the sinusoidal shape.

Mnemonics

English-language power engineering students are advised to remember: "ELI the ICE man" or "ELI on ICE" – the voltage E leads the current I in an inductor L, the current leads the voltage in a capacitor C. Another common mnemonic is CIVIL – in a capacitor (C) the current (I) leads voltage (V), voltage (V) leads current (I) in an inductor (L).

Volt-ampere Reactive

In electric power transmission and distribution, volt-ampere reactive (var) is a unit used to measure reactive power in an AC electric power system. Reactive power exists in an AC circuit when the current and voltage are not changing at the same time. The correct symbol is var and not Var, VAr, or VAR, but all three terms are widely used. The term *var* was proposed by the Romanian electrical engineer Constantin Budeanu and introduced in 1930 by the IEC in Stockholm, which has adopted it as the unit for reactive power.

Vars may be considered as either the imaginary part of apparent power, or the power flowing into a reactive load, where voltage and current are specified in volts and amperes. The two definitions are equivalent.

Reactive Power

A sinusoidally alternating voltage applied to a purely resistive load results in an alternating current that is fully in phase with the voltage. However, in many applications it is common for there to be a reactive component to the system, that is, the system possesses capacitance, inductance, or both. These electrical properties cause the current to change phase with respect to the voltage: capacitance tending the current to lead the voltage in phase, and inductance to lag it. For sinusoid currents and voltages at the same frequency, reactive power in vars is the product of the RMS voltage and current, or the apparent power, multiplied by the sine of the phase angle between the voltage and the current. The reactive power Q, (measured in units of volt-amperes reactive or var), is given by:

$$Q = V_{\text{rms}} I_{\text{rms}} \sin(\phi)$$

where ϕ is the phase angle between the voltage and current.

Only effective power, the actual power delivered to or consumed by the load, is expressed in watts. Imaginary power is properly expressed in volt-amperes reactive.

Reactive Power Control

Transmission connected generators are generally required to support reactive power flow. For example on the Great Britain transmission system generators are required by the Grid Code Requirements to supply their rated power between the limits of 0.85 power factor lagging and 0.95 power factor leading at the designated terminals. The system operator will perform switching actions to maintain a secure and economical voltage profile while maintaining a reactive power balance equation:

Generator_MVARs + System_gain + Shunt_capacitors = MVAR_Demand + Reactive_losses + Shunt_reactors

The 'System gain' is an important source of reactive power in the above power balance equation, which is generated by the capacitive nature of the transmission network itself. By making decisive switching actions in the early morning before the demand increases, the system gain can be maximised early on, helping to secure the system for the whole day. To balance the equation some pre-fault reactive generator use will be required. Other sources of reactive power that will also be used include shunt capacitors, shunt reactors, Static VAR Compensators and voltage control circuits.

Unbalanced Polyphase Systems

While real power and reactive power are well defined in any system, the definition of apparent power for unbalanced polyphase systems is considered to be one of the most controversial topics in power engineering. Major delineations of the concept are attributed to Stanley's *Phenomena of Retardation in the Induction Coil* (1888) and Steinmetz's *Theoretical Elements of Engineering* (1915). However, with the development of three phase power distribution, it became clear that the definition of apparent power and the power factor could not be applied to unbalanced polyphase systems. In 1920, a "Special Joint Committee of the AIEE and the National Electric Light Association" met to resolve the issue. They considered two definitions:

$$pf = \frac{Pa + Pb + Pc}{|Sa| + |Sb| + |Sc|}$$

that is, the quotient of the sums of the real powers for each phase over the sum of the apparent power for each phase.

$$pf = \frac{Pa + Pb + Pc}{|Pa + Pb + Pc + j(Qa + Qb + Qc)|}$$

that is, the quotient of the sums of the real powers for each phase over the magnitude of the sum of the complex powers for each phase.

The 1920 committee found no consensus and the topic continued to dominate discussions. In 1930 another committee formed and once again failed to resolve the question. The transcripts of their discussions are the lengthiest and most controversial ever published by the AIEE (Emanuel, 1993). Further resolution of this debate did not come until the late 1990s.

Basic Calculations Using Real Numbers

A perfect resistor stores no energy, so current and voltage are in phase. Therefore there is no reactive power and $P = S$. Therefore for a perfect resistor

$$P = S = V_{\text{RMS}} I_{\text{RMS}} = I_{\text{RMS}}^2 R = \frac{V_{\text{RMS}}^2}{R}$$

For a perfect capacitor or inductor there is no net power transfer, so all power is reactive. Therefore for a perfect capacitor or inductor:

P = 0

$$Q = |S| = V_{\text{RMS}} I_{\text{RMS}} = I_{\text{RMS}}^2 |X| = \frac{V_{\text{RMS}}^2}{|X|}$$

Where X is the reactance of the capacitor or inductor.

If X is defined as being positive for an inductor and negative for a capacitor then we can remove the modulus signs from Q and X and get

$$Q = I_{\text{RMS}}^2 X = \frac{V_{\text{RMS}}^2}{X}$$

Instantaneous power is defined as:

$$P_{inst} = v(t) * i(t)$$

where v(t) and i(t) are the time varying voltage and current waveforms.

This definition is useful because it applies to all waveforms, whether they are sinusoidal or not. This is particularly useful in power electronics, where nonsinusoidal waveforms are common. In general, we are interested in the real power averaged over a period of time, whether it is a low frequency line cycle or a high frequency power converter switching period. The simplest way to get that result is to take the integral of the instantaneous calculation over the desired period.

$$P_{avg} = \frac{1}{T_{period}} \int_{t_1}^{t_2} v(t) * i(t) dt.$$

This method of calculating the average power gives the real power regardless of harmonic content of the waveform. In practical applications, this would be done in the digital domain, where the calculation becomes trivial when compared to the use of rms and phase to determine real power.

$$P_{avg} = \frac{1}{T} * \sum_{k=1}^{n} v[n] * i[n]$$

Multiple Frequency Systems

Since an RMS value can be calculated for any waveform, apparent power can be calculated from this.

For real power it would at first appear that we would have to calculate loads of product terms and average all of them. However if we look at one of these product terms in more detail we come to a very interesting result.

$$\begin{aligned} A\cos(\omega_1 t + k_1)\cos(\omega_2 t + k_2) &= \frac{A}{2}\cos((\omega_1 t + k_1) + (\omega_2 t + k_2)) + \frac{A}{2}\cos((\omega_1 t + k_1) - (\omega_2 t + k_2)) \\ &= \frac{A}{2}\cos((\omega_1 + \omega_2)t + k_1 + k_2) + \frac{A}{2}\cos((\omega_1 - \omega_2)t + k_1 - k_2) \end{aligned}$$

However the time average of a function of the form cos($\omega t+k$) is zero provided that ω is nonzero. Therefore the only product terms that have a nonzero average are those where the frequency of voltage and current match. In other words it is possible to calculate real (average) power by simply treating each frequency separately and adding up the answers.

Furthermore, if we assume the voltage of the mains supply is a single frequency (which it usually is), this shows that harmonic currents are a bad thing. They will increase the rms current (since there will be non-zero terms added) and therefore apparent power, but they will have no effect on the real power transferred. Hence, harmonic currents will reduce the power factor.

Harmonic currents can be reduced by a filter placed at the input of the device. Typically this will consist of either just a capacitor (relying on parasitic resistance and inductance in the supply) or a capacitor-inductor network. An active power factor correction circuit at the input would generally reduce the harmonic currents further and maintain the power factor closer to unity.

Three-phase Electric Power

Figure : Three-phase transformer with four wire output for 208Y/120 Volt service: one wire for neutral, others for A, B and C phases.

Three-phase electric power is a common method of alternating-current electric power generation, transmission, and distribution. It is a type of polyphase system and is the most common method used by grids worldwide to transfer power. It is also used to power large motors and other heavy loads. A three-phase system is generally more economical than others because it uses less conductor material to transmit electric power than equivalent single-phase or two-phase systems at the same voltage. The three-phase system was introduced and patented by Nikola Tesla in 1887 and 1888.

In a three-phase system, three circuit conductors carry three alternating currents (of the same frequency) which reach their instantaneous peak values at different times. Taking one conductor as the reference, the other two currents are delayed in time by one-third and two-thirds of one cycle of the electric current. This delay between phases has the effect of giving constant power transfer over each cycle of the current and also makes it possible to produce a rotating magnetic field in an electric motor.

Three-phase systems may have a neutral wire. A neutral wire allows the three-phase system to use a higher voltage while still supporting lower-voltage single-phase appliances. In high-voltage distribution situations, it is common not to have a neutral wire as the loads can simply be connected between phases (phase-phase connection).

Three-phase has properties that make it very desirable in electric power systems:

- The phase currents tend to cancel out one another, summing to zero in the case of a linear balanced load. This makes it possible to eliminate or reduce the size of the neutral conductor; all the phase conductors carry the same current and so can be the same size, for a balanced load.
- Power transfer into a linear balanced load is constant, which helps to reduce generator and motor vibrations.
- Three-phase systems can produce a magnetic field that rotates in a specified direction, which simplifies the design of electric motors.

Three is the lowest phase order to exhibit all of these properties.

Most household loads are single-phase. In North America and a few other places, three-phase power generally does not enter homes. Even in areas where it does, it is typically split out at the main distribution board and the individual loads are fed from a single phase. Sometimes it is used to power electric stoves and electric clothes dryers.

The three phases are typically indicated by colours which vary by country.

Generation and Distribution

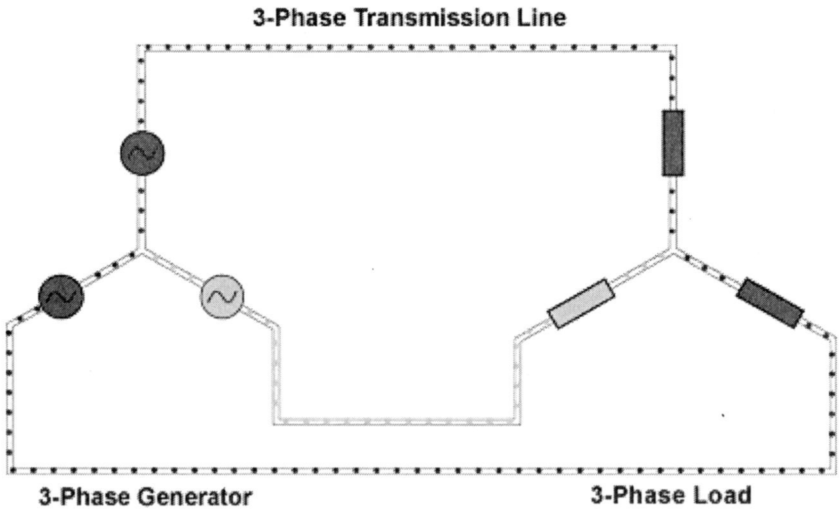

Figure : Animation of three-phase current flow

At the power station, an electrical generator converts mechanical power into a set of three AC electric currents, one from each coil (or winding) of the generator. The windings are arranged such that the currents vary sinusoidally at the same frequency but with the peaks and troughs of their wave forms offset to provide three complementary currents with a phase separation of one-third cycle (120° or D $_{3}$ radians). The generator frequency is typically 50 or 60 Hz, varying by country.

Large power generators provide an electric current at a potential which can be a few hundred volts or up to about 30 kV. At the power station, transformers step this voltage up to one suitable for transmission.

After numerous further conversions in the transmission and distribution network, the power is finally transformed to the standard utilisation voltage for lighting and equipment. Single-phase loads are connected from one phase to neutral or between two phases. Three-phase loads such as larger motors must be connected to all three phases of the supply.

Three-wire versus Four-wire

Three-phase circuits occur in two varieties. In one case, there are only three energised ("hot" or "live") wires; in the other case, there are three hot wires plus a neutral wire. Four-wire circuits offer flexibility, since a load may be connected "line-to-line" or "line-to-neutral"; three-wire circuits offer economy, since the neutral conductor is eliminated. Commonly, distribution voltage circuits are four-wire, while higher voltage transmission circuits are three-wire. Transmission lines often feature a ground wire, but this is solely for fault and lightning protection and is not connected to deliver electrical power.

Single-phase Loads

Single-phase loads may be connected to a three-phase system in two ways. Either a load may be connected across two of the live conductors, or a load can be connected from a live phase conductor to the neutral conductor. Single-phase loads should be distributed evenly between the phases of the three-phase system for efficient use of the supply transformer and supply conductors. If the line-to-neutral voltage is a standard load voltage, for example 230 volt on a 400 volt three-phase system, single-phase loads can connect to a phase and the neutral. Loads can be distributed over three phases to balance the load. Where the line-to-neutral voltage is not the standard voltage for example 347 volts produced by a 600 V system, single-phase loads are connected through a step-down transformer.

In a symmetrical three-phase system, the system neutral has the same magnitude of voltage to each of the three-phase conductors. The voltage between line conductors (V_l) is –3 times the phase conductor to neutral voltage (V_p). That is: $V_l = \cdot 3V_p$.

In some multiple-unit residential buildings of North America, three-phase power is supplied to the building but individual units have only single-phase power formed from two of the three supply phases. Lighting and convenience receptacles are connected from either phase conductor to neutral, giving the usual 120 V required by typical North American appliances. In the split-phase system, high-power loads are connected between the opposite "hot" poles, giving a voltage of 240 V. In some cases, they may be connected between phases of a three-phase system, giving a voltage of 208 V. This practice is common enough that 208 V single-phase equipment is readily available in North America. Attempts to use the more common 120/240 V equipment intended for split-phase distribution may result in poor performance since 240 V heating and lighting equipment will only produce 75% of its rating when operated at 208 V. Motors rated at 240 V will draw higher current at 208 V; some motors are dual-labelled for both voltages.

Where three-phase at low voltage is otherwise in use, it may still be split out into single-phase service cables through joints in the supply network or it may be delivered to a master distribution board (breaker panel) at the customer's premises. Connecting an electrical circuit from one phase to the neutral generally supplies the standard single phase voltage to the circuit (either 120 V AC or 230 V AC depending on the regional standard).

The currents returning from the customers' premises to the supply transformer all share the neutral wire. If the loads are evenly distributed on all three phases, the sum of the returning currents in the neutral wire is approximately zero. Any unbalanced phase loading on the secondary side of the transformer will use the transformer capacity inefficiently.

If the supply neutral of a three-phase system with line-to-neutral connected loads is broken, the voltage balance on the loads will no longer be maintained. The neutral point will tend to drift toward the most heavily loaded phase, causing undervoltage conditions on that phase and overvoltage on a lightly loaded phase; the lightly loaded phases may approach the line-to-line voltage, which exceeds the line-to-neutral voltage by a factor of –3, causing overheating and failure of many types of loads.

For example, if several houses are connected through a 240 V transformer, which is connected to one phase of the three-phase system, each house might be affected by the imbalance on the three-phase system. If the neutral connection is broken somewhere in the system, all equipment in a house might be damaged due to over-voltage. A similar phenomenon can exist if the house neutral (connected to the centre tap of the 240 V pole transformer) is disconnected. This type of failure event can be difficult to troubleshoot if the drifting neutral effect is not understood. With inductive and/or capacitive loads, all phases can suffer damage as the reactive current moves across abnormal paths in the unbalanced system, especially if resonance conditions occur. For this reason, neutral connections are a critical part of a power distribution network and must be made as reliable as any of the phase connections. Where a mixture of single-phase 120-volt lighting and three-phase, 240-volt motors are to be supplied, a system called high-leg delta is used.

Three-phase Loads

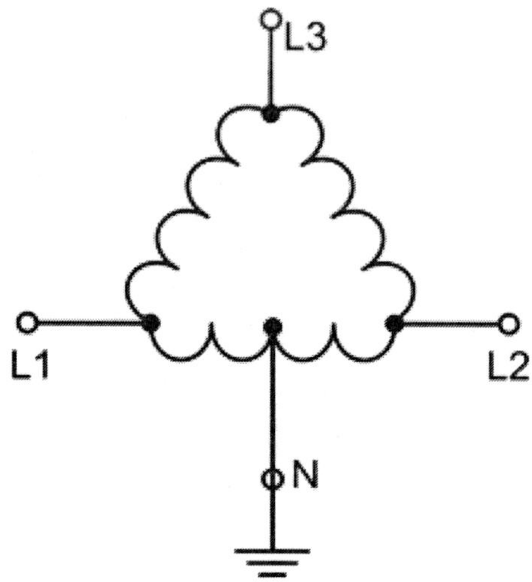

Figure : A transformer for a high-leg delta system; 240 V 3-phase motors would be connected to L1, L2 and L3. Single-phase lighting would be connected L1 or L2 to neutral (N). No loads would be connected from L3 (the high or wild leg) to neutral, since the voltage would be 208 V.

The most important class of three-phase load is the electric motor. A three-phase induction motor has a simple design, inherently high starting torque and high efficiency. Such motors are applied in industry for pumps, fans, blowers, compressors, conveyor drives, electric vehicles and many other kinds of motor-driven equipment. A three-phase motor is more compact and less costly than a single-phase motor of the same voltage class and rating and single-phase AC motors above 10 HP (7.5 kW) are uncommon. Three-phase motors also vibrate less and hence last longer than single-phase motors of the same power used under the same conditions.

Resistance heating loads such as electric boilers or space heating may be connected to three-phase systems. Electric lighting may also be similarly connected. These types of loads do not require the revolving magnetic field characteristic of three-phase motors but take advantage of the higher voltage and power level usually associated with three-phase distribution. Legacy single-phase fluorescent lighting systems also benefit from reduced flicker in a room if adjacent fixtures are powered from different phases. Large rectifier systems may have three-phase inputs; the resulting DC is easier to filter (smooth) than the output of a single-phase rectifier. Such rectifiers may be used for battery charging, electrolysis processes such as aluminium production or for operation of DC motors.

One example of a three-phase load is the electric arc furnace used in steelmaking and in refining of ores.

In much of Europe, stoves are designed for a three-phase feed. Usually the individual heating units are connected between phase and neutral to allow for connection to a single-phase supply. In many areas of Europe, single-phase power is the only source available.

Phase Converters

Occasionally the advantages of three-phase motors make it worthwhile to convert single-phase power to three-phase. Small customers, such as residential or farm properties, may not have access to a three-phase supply or may not want to pay for the extra cost of a three-phase service but may still wish to use three-phase equipment. Such converters may also allow the frequency to be varied (resynthesis) allowing speed control. Some railway locomotives are moving to multi-phase motors driven by such systems even though the incoming supply to a locomotive is nearly always either DC or single-phase AC.

Because single-phase power goes to zero at each moment that the voltage crosses zero but three-phase delivers power continuously, any such converter must have a way to store energy for the necessary fraction of a second.

One method for using three-phase equipment on a single-phase supply is with a rotary phase converter, essentially a three-phase motor with special starting arrangements and power factor correction that produces balanced three-phase voltages. When properly designed, these rotary converters can allow satisfactory operation of three-phase equipment such as machine tools on a single-phase supply. In such a device, the energy storage is performed by the mechanical inertia (flywheel effect) of the rotating components. An external flywheel is sometimes found on one or both ends of the shaft.

A second method that was popular in the 1940s and 1950s was the *transformer method.* At that time, capacitors were more expensive than transformers, so an autotransformer was used to apply more power through fewer capacitors. This method performs well and does have supporters, even today. The usage of the name *transformer method* separated it from another common method, the static converter, as both methods have no moving parts, which separates them from the rotary converters. Another method often attempted is with a device referred to as a static phase converter. This method of running three-phase equipment is commonly attempted with motor loads though it only supplies T! power and can cause the motor loads to run hot and in some cases overheat. This method does not work when sensitive circuitry is involved such as CNC devices or in induction and rectifier-type loads.

A three-phase generator can be driven by a single-phase motor. This motor-generator combination can provide a frequency changer function as well as phase conversion, but requires two machines with all their expense and losses. The motor-generator method can also form an uninterruptable power supply when used in conjunction with a large flywheel and a standby generator set.

Some devices are made which create an imitation three-phase from three-wire single-phase supplies. This is done by creating a third "subphase" between the two live conductors, resulting in a phase separation of 180° – 90° = 90°. Many three-phase devices can run on this configuration but at lower efficiency. Variable-frequency drives (also known as solid-state inverters) are used to provide precise speed and torque control of three-phase motors. Some models can be powered by a single-phase supply. VFDs work by converting the supply voltage to DC and then converting the DC to a suitable three-phase source for the motor.

Digital phase converters are designed for fixed-frequency operation from a single-phase source. Similar to a variable-frequency drive, they use a microprocessor to control solid-state power switching components to maintain balanced three-phase voltages.

Alternatives to Three-phase

- Split-phase electric power is used when three-phase power is not available and allows double the normal utilisation voltage to be supplied for high-power loads.
- Two-phase electric power, like three-phase, gives constant power transfer to a linear load. For loads that connect each phase to neutral, assuming the load is the same power draw, the two-wire system has a neutral current which is greater than neutral current in a three-phase system. Also motors are not entirely linear, which means that despite the theory, motors running on three-phase tend to run smoother than those on two-phase. The generators in the Adams Power Plant at Niagara Falls which were installed in 1895 were the largest generators in the world at the time and were two-phase machines. True two-phase power distribution is basically obsolete. Special-purpose systems may use a two-phase system for control. Two-phase power may be obtained from a three-phase system (or

vice versa) using an arrangement of transformers called a Scott-T transformer.

- *Monocyclic power* was a name for an asymmetrical modified two-phase power system used by General Electric around 1897, championed by Charles Proteus Steinmetz and Elihu Thomson. This system was devised to avoid patent infringement. In this system, a generator was wound with a full-voltage single-phase winding intended for lighting loads and with a small fraction (usually ¼ of the line voltage) winding which produced a voltage in quadrature with the main windings. The intention was to use this "power wire" additional winding to provide starting torque for induction motors, with the main winding providing power for lighting loads. After the expiration of the Westinghouse patents on symmetrical two-phase and three-phase power distribution systems, the monocyclic system fell out of use; it was difficult to analyse and did not last long enough for satisfactory energy metering to be developed.
- High-phase-order systems for power transmission have been built and tested. Such transmission lines use six (two-pole, three-phase) or twelve (two-pole, six-phase) lines and employ design practices characteristic of extra-high-voltage transmission lines. High-phase-order transmission lines may allow transfer of more power through a given transmission line right-of-way without the expense of a high-voltage direct current (HVDC) converter at each end of the line.

Three-phase AC Railway Electrification

Three-phase AC railway electrification was used in Italy, Switzerland and the United States in the early twentieth century. Italy was the major user, from 1901 until the 1976, although lines through two tunnels also used the system; the Simplon Tunnel in Switzerland from 1906 to 1930, and the Cascade Tunnel of the Great Northern Railway in the United States from 1909 to 1939. The first line was in Switzerland, from Burgdorf to Thun (40 km, 25 mi), since 1899.

The system provides regenerative braking with the power fed back to the system, so is particularly suitable for mountain railways (provided another locomotive on the line can use the power). The locomotives use three-phase induction motors without commutators which require less maintenance. The early Italian and Swiss sysyems used a low frequency (16T! Hz), and a relatively low voltage (3,000 or 3,600 volts) compared with later AC systems. The overhead wiring, generally having two separate overhead lines and the rail for the third phase, was more complicated, and the low-frequency used required a separate generation or conversion and distribution system. Train speed was restricted to one to four speeds, with two or four speeds obtained by pole-changing or cascade operation or both.

The system is only used today for four mountain (funicular or rack) railways using powered carriages, where the overhead wiring is less complicated and restrictions on the speeds available less important; the Corcovado Rack Railway in Rio de Janeiro Brazil, Gornergratbahn and Jungfraubahn in Switzerland and the Petit train de la Rhune in France). They are industrial rather than low frequency (50 Hz, or 60 Hz (Brazil)), using between 725 and 3000 volts.

This category does not cover railways with a single-phase (or DC) supply which is converted to three-phase on the locomotive or power car, e.g. most railway equipment from the 1990s and earlier using solid-state converters. The Kando system of the 1930s developed by Kando and used in Hungary and Italy used rotating converters on the locomotive to convert the single-phase supply to three phases, as did the Phase-splitting system on the Norfolk and Western Railroad in the USA.

Locomotives

Usually the locomotives had one, two or four motors on the body chassis (not on the bogies), and did not require gearing. The induction motors are designed to run at a particular synchronous speed, and when they run above the synchronous speed downhill, power is fed back to the system. Pole changing and cascade (concantation) working was used to allow two or four different speeds, and resistances (often liquid rheostats) were required for starting. In Italy freight locomotives used plain cascade with two speeds, 25 and 50 km/h (16 and 31 mph); while express locomotives used cascade combined with pole-changing giving four speeds, 37, 50, 75 and 100 km/h (23, 31, 46 and 62 mph). With the use of 3,000 or 3,600 volts at 16T! (16.7) Hz, the supply could be fed directly to the motor, so not requiring an onboard transformer.

Generally the motor(s) fed a single axle with other wheels linked by connecting rods, as the induction motor is sensitive to speed variations and with non-linked motors on several axles the motors on worn wheels would do little or even no work as they would rotate faster. This motor charactership led to a mishap in the Cascade Tunnel to a GN east-bound freight train with four electric locos, two on the head and two pushing. The two pushers suddenly lost power and the train gradually slowed to a stop.

But the lead unit engineer was unaware that his train had stopped, and held the controller on the power position until the usual time to transit the tunnel had elapsed. Not seeing daylight, he finally shut down the locomotive, and found that the wheels of his stationary loco had ground through two-thirds of the rail web.

Overhead Wiring

Generally two separate overhead wires are used, with the rail for the third phase, though occasionally three overhead wires are used. At junctions, crossovers and crossings the two lines must be kept apart, with a continuous supply to the locomotive, which must have two live conductors wherever it stops.

Hence two collectors per overhead phase are used, but the possibility of bridging a dead section and causing a short circuit from the front collector of one phase to the back collector of the other phase must be avoided. The resistance of the rails used for the third phase or return is higher for AC than for DC because of the "skin" effect, but lower for the low frequency used than for industrial frequency. Losses are also increased, though not in the same proportion, as the impedance is largely reactive.

The locomotive needs to pick up power from two (or three) overhead conductors. Early locomotives on the Italian State Railways used a wide bow collector which covered both wires but later locomotives used two pantographs side by side. In the United States, a pair of trolley poles were used. They worked well with a maximum speed of 15 miles per hour (24 km/h). The dual conductor pantograph system is used on four mountain railways that continue to use three-phase power (Corcovado Rack Railway in Rio de Janeiro, Brazil, Jungfraubahn and Gornergratbahn in Switzerland and the Petit train de la Rhune in France).

Per Unit Representation

In a power system different power equipment with different voltage and power levels are connected together through various step up or step down transformers. However the presence of various voltage and power levels causes problem in finding out the currents (or voltages) at different points in the network.

To alleviate this problem, all the system quantities are converted into a uniform normalised platform. This is called the per unit system.

In a per unit system each system variable or quantity is normalised with respect to its own base value. The units of these normalised values are per unit (abbreviated as pu) and not Volt, Ampere or Ohm. The base quantities chosen are:

- VA base (P_{base}): This is the three-phase apparent power (Volt-Ampere) base that is common to the entire circuit.
- Voltage Base (V_{base}): This is the line-to-line base voltage. This quantity is not uniform for the entire circuit but gets changed by the turns ratio of the transformer.

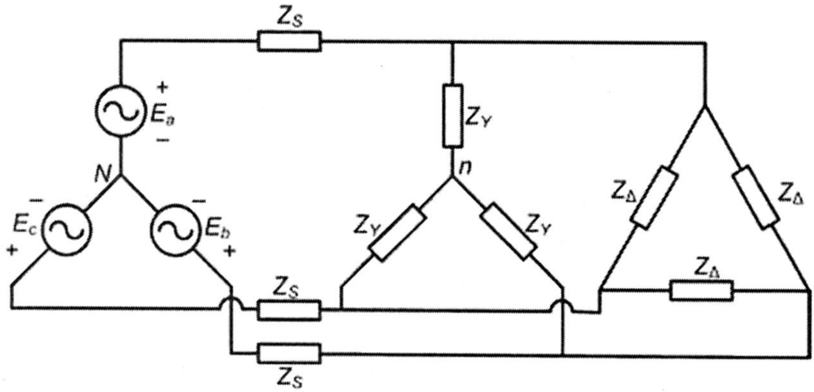

Figure : Three balanced sources supplying two balanced load through balanced source impedances.

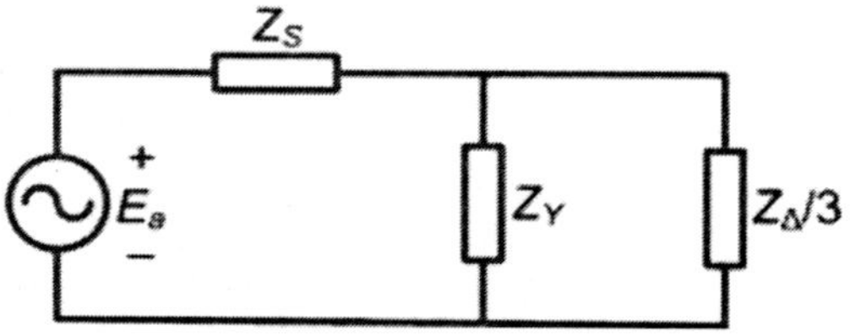

Figure : Per phase equivalent circuit of the network

Automatic Gain Control

Automatic gain control (AGC) is an adaptive system found in many electronic devices. The average output signal level is fed back to adjust the gain to an appropriate level for a range of input signal levels. For example, without AGC the sound emitted from an AM radio receiver would vary to an extreme extent from a weak to a strong signal; the AGC effectively reduces the volume if the signal is strong and raises it when it is weaker.

Example use Cases

AM Radio

In 1925, Harold Alden Wheeler invented automatic volume control (AVC) and obtained a patent. Karl Küpfmüller published an analysis of AGC systems in 1928. By the early 1930s essentially all broadcast receivers included automatic volume control.

AGC is a departure from linearity in AM radio receivers. Without AGC, an AM radio would have a linear relationship between the signal amplitude and the sound waveform — the sound amplitude, which correlates with loudness, is proportional to the radio signal amplitude, because the information content of the signal is carried by the changes of amplitude of the carrier wave. If the circuit were not fairly linear, the modulation could not be recovered with reasonable fidelity. However, the strength of the signal received will vary widely, depending on the power and distance of the transmitter, and signal path attenuation.

The AGC circuit keeps the receiver's output level from fluctuating too much by detecting the overall strength of the signal and automatically adjusting the gain of the receiver to maintain an approximately constant average output level. For a very weak signal, the AGC has no effect, allowing the receiver to operate at its maximum gain; as the signal increases, the AGC reduces the gain.

It is usually disadvantageous to reduce the gain of the RF front end of the receiver on weaker signals as low gain can worsen signal-to-noise ratio and blocking; therefore, many designs reduce gain only for stronger signals.

Since the AM detector diode produces a DC voltage proportional to signal strength, this voltage can be fed back to earlier stages of the receiver to reduce gain. A filter network is required so that the audio components of the signal don't appreciably influence gain; this prevents "modulation rise" which increases the effective modulation depth of the signal, distorting the sound. Communications receivers may have more complex AVC systems, including extra amplification stages, separate AGC detector diodes, different time constants for broadcast and shortwave bands, and application of different levels of AGC voltage to different stages of the receiver to prevent distortion and cross-modulation. Design of the AVC system has a great effect on the usability of the receiver, tuning characteristics, audio fidelity, and behaviour on overload and strong signals.

FM receivers, even though they incorporate limiter stages and detectors that are relatively insensitive to amplitude variations, still benefit from AGC to prevent overload on strong signals.

Radar

A related application of AGC is in radar systems, as a method of overcoming unwanted clutter echoes. This method relies on the fact that clutter returns far outnumber echoes from targets of interest. The receiver's gain is automatically adjusted to maintain a constant level of overall visible clutter. While this does not help detect targets masked by stronger surrounding clutter, it does help to distinguish strong target sources. In the past, radar AGC was electronically controlled and affected the gain of the entire radar receiver. As radars evolved, AGC became computer-software controlled, and affected the gain with greater granularity, in specific detection cells.

Audio/Video

An audio tape generates a certain amount of noise. If the level of the signal on the tape is low, the noise is more prominent, i.e., the signal-to-noise ratio is lower than it could be. To produce the least noisy recording, the recording level should be set as high as possible without being so high as to clip or seriously distort the signal. In professional high-fidelity recording the level is set manually using a peak-reading meter. If high fidelity is not a requirement, a suitable recording level can be set by an AGC circuit which reduces the gain as the average signal level increases. This allows a usable recording to be made even for speech some distance from the microphone of an audio recorder. Similar considerations apply with VCRs.

A potential disadvantage of AGC is that when recording something like music with quiet and loud passages such as classical music, the AGC will tend to make the quiet passages louder and the loud passages quieter, compressing the dynamic range; the result can be a reduced musical quality if the signal is not re-expanded when playing, as in a companding system.

Most reel-to-reel tape recorders and cassette decks have AGC circuits. Those used for high-fidelity allow it to be overridden manually.

Most VCR circuits use the amplitude of the vertical blanking pulse to operate the AGC. Video copy control schemes such as Macrovision exploit this, inserting spikes in the pulse which will be ignored by most television sets, but cause a VCR's AGC to overcorrect and corrupt the recording.

Telephone Recording

Devices to record both sides of a telephone conversation must record both the relatively large signal from the local user and the much smaller signal from the remote user at comparable loudnesses. Some telephone recording devices incorporate automatic gain control to produce acceptable-quality recordings.

Biological

As is the case with many concepts found in engineering, automatic gain control is also found in biological systems, especially sensory systems. For example, in the vertebrate visual system, calcium dynamics in the retinal photoreceptors adjust gain to suit light levels. Further on in the visual system, cells in V1 are thought to mutually inhibit, causing normalisation of responses to contrast, a form of automatic gain control. Similarly, in the auditory system, the olivocochlear efferent neurons are part of a bio-mechanical gain control loop.

Chapter 3

Power Engineering

Power engineering, also called power systems engineering, is a subfield of engineering that deals with the generation, transmission and distribution of electric power as well as the electrical devices connected to such systems including generators, motors and transformers. Although much of the field is concerned with the problems of three-phase AC power-the standard for large-scale power transmission and distribution across the modern world-a significant fraction of the field is concerned with the conversion between AC and DC power as well as the development of specialised power systems such as those used in aircraft or for electric railway networks.

History

Electricity became a subject of scientific interest in the late 17th century with the work of William Gilbert. Over the next two centuries a number of important discoveries were made including the incandescent lightbulb and the voltaic pile. Probably the greatest discovery with respect to power engineering came from Michael Faraday who in 1831 discovered that a change in magnetic flux induces an electromotive force in a loop of wire—a principle known as electromagnetic induction that helps explain how generators and transformers work.

In 1881 two electricians built the world's first power station at Godalming in England. The station employed two waterwheels to produce an alternating current that was used to supply seven Siemens arc lamps at 250 volts and thirty-four incandescent lamps at 40 volts. However supply was intermittent and in 1882 Thomas Edison and his company, The Edison Electric Light Company, developed the first steam-powered electric power station on Pearl Street in New York City. The Pearl Street Station consisted of several generators and initially powered around 3,000 lamps for 59 customers. The power station used direct current and operated at a single voltage. Since the direct current power could not be easily transformed to the higher voltages necessary to minimise power loss during transmission, the possible distance between the generators and load was limited to around half-a-mile (800 m).

Figure : A sketch of the Pearl Street Station

That same year in London Lucien Gaulard and John Dixon Gibbs demonstrated the first transformer suitable for use in a real power system. The practical value of Gaulard and Gibbs' transformer was demonstrated in 1884 at Turin where the transformer was used to light up forty kilometres (25 miles) of railway from a single alternating current generator. Despite the success of the system, the pair made some fundamental mistakes.

Perhaps the most serious was connecting the primaries of the transformers in series so that switching one lamp on or off would affect other lamps further down the line. Following the demonstration George Westinghouse, an American entrepreneur, imported a number of the transformers along with a Siemens generator and set his engineers to experimenting with them in the hopes of improving them for use in a commercial power system.

One of Westinghouse's engineers, William Stanley, recognised the problem with connecting transformers in series as opposed to parallel and also realised that making the iron core of a transformer a fully enclosed loop would improve the voltage regulation of the secondary winding. Using this knowledge he built a much improved alternating current power system at Great Barrington, Massachusetts in 1886. Then in 1887 and 1888 another engineer called Nikola Tesla filed a range of patents related to power systems including one for a two-phase induction motor. Although Tesla cannot necessarily be attributed with building the first induction motor, his design, unlike others, was practical for industrial use.

By 1890 the power industry had flourished and power companies had built literally thousands of power systems (both direct and alternating current) in the United States and Europe-these networks were effectively dedicated to providing electric lighting. During this time a fierce rivalry known as the "War of Currents" emerged between Edison, Westinghouse and Tesla over which form of transmission (direct or alternating current) was superior. In 1891, Westinghouse installed the first major power system that was designed to drive an electric motor and not just provide electric lighting. The installation powered a 100 horsepower (75 kW) synchronous motor at Telluride, Colorado with the motor being started by a Tesla induction motor. On the other side of the Atlantic, Oskar von Miller built a 20 kV 176 km three-phase transmission line from Lauffen am Neckar to Frankfurt am Main for the Electrical Engineering Exhibition in Frankfurt. In 1895, after a protracted decision-making process, the Adams No. 1 generating station at Niagara Falls began transmitting three-phase alternating current power to Buffalo at 11 kV. Following completion of the Niagara Falls project, new power systems increasingly chose alternating current as opposed to direct current for electrical transmission.

Although the 1880s and 1890s were seminal decades in the field, developments in power engineering continued throughout the 20th and 21st century. In 1936 the first commercial HVDC (high voltage direct current) line using Mercury arc valves was built between Schenectady and Mechanicville, New York. HVDC had previously been achieved by installing direct current generators in series (a system known as the Thury system) although this suffered from serious reliability issues. In 1957 Siemens demonstrated the first solid-state rectifier (solid-state rectifiers are now the standard for HVDC systems) however it was not until the early 1970s that this technology was used in commercial power systems. In 1959 Westinghouse demonstrated the first circuit breaker that used SF_6 as the interrupting medium. SF_6 is a far superior dielectric to air and, in recent times, its use has been extended to produce far more compact switching equipment (known as switchgear) and transformers. Many important developments also came from extending innovations in the information technology and telecommunications field to the power engineering field. For example, the development of computers meant load flow studies could be run more efficiently allowing for much better planning of power systems. Advances in information technology and telecommunication also allowed for much better remote control of the power system's switchgear and generators.

Basics of Electric Power

Electric power is the mathematical product of two quantities: current and voltage. These two quantities can vary with respect to time (AC power) or can be kept at constant levels (DC power).

Most refrigerators, air conditioners, pumps and industrial machinery use AC power whereas most computers and digital equipment use DC power (the digital devices you plug into the mains typically have an internal or external power adapter to convert from AC to DC power). AC power has the advantage of being easy to transform between voltages and is able to be generated and utilised by brushless machinery. DC power remains the only practical choice in digital systems and can be more economical to transmit over long distances at very high voltages. The ability to easily transform the voltage of AC power is important for two reasons: Firstly, power can be transmitted over long distances with less loss at higher voltages. So in power networks where generation is distant from the load, it is desirable to step-up the voltage of power at the generation point and then step-down the voltage near the load. Secondly, it is often more economical to install turbines that produce higher voltages than would be used by most appliances, so the ability to easily transform voltages means this mismatch between voltages can be easily managed. Solid state devices, which are products of the semiconductor revolution, make it possible to transform DC power to different voltages, build brushless DC machines and convert between AC and DC power. Nevertheless devices utilising solid state technology are often more expensive than their traditional counterparts, so AC power remains in widespread use.

Power

Power Engineering deals with the generation, transmission and distribution of electricity as well as the design of a range of related devices. These include transformers, electric generators, electric motors and power electronics.

The power grid is an electrical network that connects a variety of electric generators to the users of electric power. Users purchase electricity from the grid avoiding the costly exercise of having to generate their own. Power engineers may work on the design and maintenance of the power grid as well as the power systems that connect to it. Such systems are called on-grid power systems and may supply the grid with additional power, draw power from the grid or do both.

Power engineers may also work on systems that do not connect to the grid. These systems are called off-grid power systems and may be used in preference to on-grid systems for a variety of reasons. For example, in remote locations it may be cheaper for a mine to generate its own power rather than pay for connection to the grid and in most mobile applications connection to the grid is simply not practical.

Today, most grids adopt three-phase electric power with alternating current. This choice can be partly attributed to the ease with which this type of power can be generated, transformed and used. Often (especially in the USA), the power is split before it reaches residential customers whose low-power appliances rely upon single-phase electric power. However, many larger industries and organisations still prefer to receive the three-phase power directly because it can be used to drive highly efficient electric motors such as three-phase induction motors.

Transformers play an important role in power transmission because they allow power to be converted to and from higher voltages. This is important because higher voltages suffer less power loss during transmission. This is because higher voltages allow for lower current to deliver the same amount of power, as power is the product of the two. Thus, as the voltage steps up, the current steps down. It is the current flowing through the components that result in both the losses and the subsequent heating. These losses, appearing in the form of heat, are equal to the current squared times the electrical resistance through which the current flows, so as the voltage goes up the losses are dramatically reduced.

For these reasons, electrical substations exist throughout power grids to convert power to higher voltages before transmission and to lower voltages suitable for appliances after transmission.

Components

Power engineering is a network of interconnected components which convert different forms of energy to electrical energy. Modern power engineering consists of three main subsystems: the generation subsystem, the transmission subsystem, and the distribution subsystem. In the generation subsystem, the power plant produces the electricity. The transmission subsystem transmits the electricity to the load centres. The distribution subsystem continues to transmit the power to the customers.

Generation

Generation of electrical power is a process whereby energy is transformed into an electrical form. There are several different transformation processes, among which are chemical, photo-voltaic, and electromechanical. Electromechanical energy conversion is used in converting energy from coal, petroleum, natural gas, uranium into electrical energy. Of these, all except the wind energy conversion process take advantage of the synchronous AC generator coupled to a steam, gas or hydro turbine such that the turbine converts steam, gas, or water flow into rotational energy, and the synchronous generator then converts the rotational energy of the turbine into electrical energy. It is the turbine-generator conversion process that is by far most economical and consequently most common in the industry today.

The AC synchronous machine is the most common technology for generating electrical energy. It is called synchronous because the composite magnetic field produced by the three stator windings rotate at the same speed as the magnetic field produced by the field winding on the rotor. A simplified circuit model is used to analyse steady-state operating conditions for a synchronous machine. The phasor diagram is an effective tool for visualising the relationships between internal voltage, armature current, and terminal voltage. The excitation control system is used on synchronous machines to regulate terminal voltage, and the turbine-governor system is used to regulate the speed of the machine. However, in highly interconnected systems, such as the "Western system", the "Texas system" and the "Eastern system", *one* machine will usually be assigned as the so-called "swing machine", and which generation may be increased or decreased to compensate for small changes in load, thereby maintaining the system frequency at precisely 60 Hz. Should the load dramatically change, as which happens with a system separation, then a combination of "spinning reserve" and the "swing machine" may be used by the system's load dispatcher.

The operating costs of generating electrical energy is determined by the fuel cost and the efficiency of the power station. The efficiency depends on generation level and can be obtained from the heat rate curve. We may also obtain the incremental cost curve from the heat rate curve. *Economic dispatch* is the process of allocating the required load demand between the available generation units such that the cost of operation is minimised. *Emission dispatch* is the process of allocating the required load demand between the available generation units such that air pollution occurring from operation is minimised. In large systems, particularly in the West, a combination of economic and emission dispatch may be used.

Transmission

The electricity is transported to load locations from a power station to a transmission subsystem. Therefore we may think of the transmission system as providing the medium of transportation for electric energy. The transmission system may be subdivided into the bulk transmission system and the sub-transmission system. The functions of the bulk transmission are to interconnect generators, to interconnect various areas of the network, and to transfer electrical energy from the generators to the major load centres. This portion of the system is called "bulk" because it delivers energy only to so-called bulk loads such as the distribution system of a town, city, or large industrial plant. The function of the sub-transmission system is to interconnect the bulk power system with the distribution system.

Transmission circuits may be built either underground or overhead. Underground cables are used predominantly in urban areas where acquisition of overhead rights of way are costly or not possible. They are also used for transmission under rivers, lakes and bays. Overhead transmission is used otherwise because, for a given voltage level, overhead conductors are much less expensive than underground cables.

The transmission system is a highly integrated system. It is referred to the substation equipment and transmission lines. The substation equipment contain the transformers, relays, and circuit breakers. Transformers are important static devices which transfer electrical energy from one circuit with another in the transmission subsystem. Transformers are used to step up the voltage on the transmission line to reduce the power loss which is dissipated on the way. A relay is functionally a level-detector; they perform a switching action when the input voltage (or current) meets or exceeds a specific and adjustable value. A circuit breaker is an automatically operated electrical switch designed to protect an electrical circuit from damage caused by overload or short circuit. A change in the status of any one component can significantly affect the operation of the entire system. Without adequate contact protection, the occurrence of undesired electric arcing causes significant degradation of the contacts, which suffer serious damage. There are three possible causes for power flow limitations to a transmission line. These causes are thermal overload, voltage instability, and rotor angle instability. Thermal overload is caused by excessive current flow in a circuit causing overheating. Voltage instability is said to occur when the power required to maintain voltages at or above acceptable levels exceeds the available power. Rotor angle instability is a dynamic problem that may occur following faults, such as short circuit, in the transmission system. It may also occur tens of seconds after a fault due to poorly damped or undamped oscillatory response of the rotor motion. As long as the *equal area criteria* is maintained, the interconnected system will remain stable. Should the *equal area criteria* be violated, it becomes necessary to separate the unstable component from the remainder of the system.

Distribution

The distribution system transports the power from the transmission system to the customer. The distribution systems are typically radial because networked systems are more expensive. The equipment associated with the distribution system includes the substation transformers connected to the transmission systems, the distribution lines from the transformers to the customers and the protection and control equipment between the transformer and the customer. The protection equipment includes lightning protectors, circuit breakers, disconnectors and fuses. The control equipment includes voltage regulators, capacitors, relays and demand side management equipment.

Power System Protection

Power system protection is a branch of electrical power engineering that deals with the protection of electrical power systems from faults through the isolation of faulted parts from the rest of the electrical network. The objective of a protection scheme is to keep the power system stable by isolating only the components that are under fault, whilst leaving as much of the network as possible still in operation. Thus, protection schemes must apply a very pragmatic and pessimistic approach to clearing system faults. For this reason, the technology and philosophies utilised in protection schemes can often be old and well-established because they must be very reliable

Components

Protection systems usually comprise five components:

- Current and voltage transformers to step down the high voltages and currents of the electrical power system to convenient levels for the relays to deal with;
- Protective relays to sense the fault and initiate a trip, or disconnection, order;
- Circuit breakers to open/close the system based on relay and autorecloser commands;
- Batteries to provide power in case of power disconnection in the system.
- Communication channels to allow analysis of current and voltage at remote terminals of a line and to allow remote tripping of equipment.

For parts of a distribution system, fuses are capable of both sensing and disconnecting faults. Failures may occur in each part, such as insulation failure, fallen or broken transmission lines, incorrect operation of circuit breakers, short circuits and open circuits. Protection devices are installed with the aims of protection of assets, and ensure continued supply of energy. The three classes of protective devices are:

Circuit Breaker

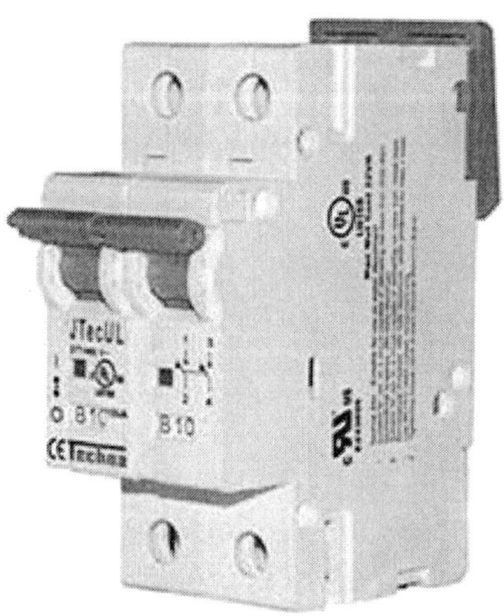

Figure : An air circuit breaker for low voltage (less than 1000 volts) power distribution switchgear

A circuit breaker is an automatically operated electrical switch designed to protect an electrical circuit from damage caused by overload or short circuit. Its basic function is to detect a fault condition and, by interrupting continuity, to immediately discontinue electrical flow. Unlike a fuse, which operates once and then has to be replaced, a circuit breaker can be reset (either manually or automatically) to resume normal operation. Circuit breakers are made in varying sizes, from small devices that protect an individual household appliance up to large switchgear designed to protect high voltage circuits feeding an entire city.

Origins

Inspired by the works of American scientist Joseph Henry and English scientist Michael Faraday, the circuit breaker was invented in 1836 by an American, Charles Grafton Page.

An early form of circuit breaker was described by Thomas Alva Edison in an 1879 patent application, although his commercial power distribution system used fuses. Its purpose was to protect lighting circuit wiring from accidental short-circuits and overloads. A modern miniature circuit breaker similar to the ones now in use was patented by Brown, Boveri & Cie in 1924. Hugo Stotz, an engineer who had sold his company, Stotz-Kontakt, to BBC, was credited as the inventor on DRP (*Deutsches Reichspatent*) 458329. Stotz's invention was the forerunner of the modern thermal-magnetic breaker commonly used in household load centres to this day.

Operation

All circuit breakers have common features in their operation, although details vary substantially depending on the voltage class, current rating and type of the circuit breaker.

The circuit breaker must detect a fault condition; in low-voltage circuit breakers this is usually done within the breaker enclosure. Circuit breakers for large currents or high voltages are usually arranged with pilot devices to sense a fault current and to operate the trip opening mechanism. The trip solenoid that releases the latch is usually energised by a separate battery, although some high-voltage circuit breakers are self-contained with current transformers, protection relays, and an internal control power source.

Once a fault is detected, contacts within the circuit breaker must open to interrupt the circuit; some mechanically-stored energy (using something such as springs or compressed air) contained within the breaker is used to separate the contacts, although some of the energy required may be obtained from the fault current itself. Small circuit breakers may be manually operated; larger units have solenoids to trip the mechanism, and electric motors to restore energy to the springs.

The circuit breaker contacts must carry the load current without excessive heating, and must also withstand the heat of the arc produced when interrupting (opening) the circuit. Contacts are made of copper or copper alloys, silver alloys, and other highly conductive materials. Service life of the contacts is limited by the erosion of contact material due to arcing while interrupting the current. Miniature and moulded case circuit breakers are usually discarded when the contacts have worn, but power circuit breakers and high-voltage circuit breakers have replaceable contacts. When a current is interrupted, an arc is generated. This arc must be contained, cooled, and extinguished in a controlled way, so that the gap between the contacts can again withstand the voltage in the circuit. Different circuit breakers use vacuum, air, insulating gas, or oil as the medium in which the arc forms. Different techniques are used to extinguish the arc including:

- Lengthening / deflection of the arc
- Intensive cooling (in jet chambers)
- Division into partial arcs
- Zero point quenching (Contacts open at the zero current time crossing of the AC waveform, effectively breaking no load current at the time of opening. The zero crossing occurs at twice the line frequency i.e. 100 times per second for 50 Hz and 120 times per second for 60 Hz AC)
- Connecting capacitors in parallel with contacts in DC circuits

Finally, once the fault condition has been cleared, the contacts must again be closed to restore power to the interrupted circuit.

Arc Interruption

Miniature low-voltage circuit breakers use air alone to extinguish the arc. Larger ratings will have metal plates or non-metallic arc chutes to divide and cool the arc. Magnetic blowout coils or permanent magnets deflect the arc into the arc chute.

In larger ratings, oil circuit breakers rely upon vapourisation of some of the oil to blast a jet of oil through the arc.

Gas (usually sulphur hexafluoride) circuit breakers sometimes stretch the arc using a magnetic field, and then rely upon the dielectric strength of the sulphur hexafluoride (SF_6) to quench the stretched arc.

Vacuum circuit breakers have minimal arcing (as there is nothing to ionise other than the contact material), so the arc quenches when it is stretched a very small amount (<2–3 mm). Vacuum circuit breakers are frequently used in modern medium-voltage switchgear to 35,000 volts.

Air circuit breakers may use compressed air to blow out the arc, or alternatively, the contacts are rapidly swung into a small sealed chamber, the escaping of the displaced air thus blowing out the arc.

Circuit breakers are usually able to terminate all current very quickly: typically the arc is extinguished between 30 ms and 150 ms after the mechanism has been tripped, depending upon age and construction of the device.

Short-circuit Current

Circuit breakers are rated both by the normal current that they are expected to carry, and the maximum short-circuit current that they can safely interrupt.

Under short-circuit conditions, a current many times greater than normal can exist. When electrical contacts open to interrupt a large current, there is a tendency for an arc to form between the opened contacts, which would allow the current to continue. This condition can create conductive ionised gases and molten or vapourised metal which can cause further continuation of the arc, or creation of additional short circuits, potentially resulting in the explosion of the circuit breaker and the equipment that it is installed in. Therefore, circuit breakers must incorporate various features to divide and extinguish the arc.

In air-insulated and miniature breakers an *arc chute* structure consisting (often) of metal plates or ceramic ridges cools the arc, and magnetic blowout coils deflect the arc into the arc chute. Larger circuit breakers such as those used in electrical power distribution may use vacuum, an inert gas such as sulphur hexafluoride or have contacts immersed in oil to suppress the arc.

The maximum short-circuit current that a breaker can interrupt is determined by testing. Application of a breaker in a circuit with a prospective short-circuit current higher than the breaker's interrupting capacity rating may result in failure of the breaker to safely interrupt a fault. In a worst-case scenario the breaker may successfully interrupt the fault, only to explode when reset.

Miniature circuit breakers used to protect control circuits or small appliances may not have sufficient interrupting capacity to use at a panelboard; these circuit breakers are called "supplemental circuit protectors" to distinguish them from distribution-type circuit breakers.

Standard Current Ratings

International Standard IEC 60898-1 and European Standard EN 60898-1 define the *rated current* I_n of a circuit breaker for low voltage distribution applications as the maximum current that the breaker is designed to carry continuously (at an ambient air temperature of 30 °C). The commonly-available preferred values for the rated current are 6 A, 10 A, 13 A, 16 A, 20 A, 25 A, 32 A, 40 A, 50 A, 63 A, 80 A, 100 A, and 125A (Renard series, slightly modified to include current limit of British BS 1363 sockets). The circuit breaker is labelled with the rated current in amperes, but without the unit symbol "A". Instead, the ampere figures is preceded by a letter "B", "C" or "D" that indicates the *instantaneous tripping current*, that is the minimum value of current that causes the circuit-breaker to trip without intentional time delay (i.e., in less than 100 ms), expressed in terms of I_n:

Type	*Instantaneous Tripping Current*
B	above 3 I_n up to and including 5 I_n
C	above 5 I_n up to and including 10 I_n
D	above 10 I_n up to and including 20 I_n
K	above 8 I_n up to and including 12 I_n For the protection of loads that cause frequent short duration (approximately 400 ms to 2 s) current peaks in normal operation.
Z	above 2 I_n up to and including 3 I_n for periods in the order of tens of seconds. For the protection of loads such as semiconductor devices or measuring circuits using current transformers.

In the United States, Underwriters Laboratories (UL) certifies equipment ratings, called Series Ratings (or "integrated equipment ratings"), using a two-tier rating. For example, a 22/10 rating. This rating means that the meter pack has a 22 kAIC tenant breaker, feeding a 10 kAIC loadcentre with 10 kAIC branches, where kAIC stands for "Thousand Ampere Interrupting Capacity." Common meter pack ratings are 22/10, 42/10 and 100/10.

Types of Circuit Breakers

Many different classifications of circuit breakers can be made, based on their features such as voltage class, construction type, interrupting type, and structural features.

Low Voltage Circuit Breakers

Low voltage (less than 1000 V_{AC}) types are common in domestic, commercial and industrial application, and include:

- MCB (Miniature Circuit Breaker)—rated current not more than 100 A. Trip characteristics normally not adjustable. Thermal or thermal-magnetic operation. Breakers illustrated above are in this category.
- MCCB (Moulded Case Circuit Breaker)—rated current up to 2500 A. Thermal or thermal-magnetic operation. Trip current may be adjustable in larger ratings.
- Low voltage power circuit breakers can be mounted in multi-tiers in low-voltage switchboards or switchgear cabinets.

The characteristics of Low Voltage circuit breakers are given by international standards such as IEC 947. These circuit breakers are often installed in draw-out enclosures that allow removal and interchange without dismantling the switchgear.

Large low-voltage moulded case and power circuit breakers may have electrical motor operators, allowing them to be tripped (opened) and closed under remote control. These may form part of an automatic transfer switch system for standby power.

Low-voltage circuit breakers are also made for direct-current (DC) applications, for example DC supplied for subway lines. Special breakers are required for direct current because the arc does not have a natural tendency to go out on each half cycle as for alternating current. A direct current circuit breaker will have blow-out coils which generate a magnetic field that rapidly stretches the arc when interrupting direct current.

Small circuit breakers are either installed directly in equipment, or are arranged in a breaker panel.

The 10 ampere DIN rail-mounted thermal-magnetic miniature circuit breaker is the most common style in modern domestic consumer units and commercial electrical distribution boards throughout Europe. The design includes the following components:

1. Actuator lever-used to manually trip and reset the circuit breaker. Also indicates the status of the circuit breaker (On or Off/tripped). Most breakers are designed so they can still trip even if the lever is held or locked in the "on" position. This is sometimes referred to as "free trip" or "positive trip" operation.
2. Actuator mechanism-forces the contacts together or apart.

3. Contacts-Allow current when touching and break the current when moved apart.
4. Terminals
5. Bimetallic strip.
6. Calibration screw-allows the manufacturer to precisely adjust the trip current of the device after assembly.
7. Solenoid
8. Arc divider/extinguisher.

Figure : Photo of inside of a circuit breaker

Magnetic Circuit Breakers

Magnetic circuit breakers use a solenoid (electromagnet) whose pulling force increases with the current. Certain designs utilise electromagnetic forces in addition to those of the solenoid. The circuit breaker contacts are held closed by a latch. As the current in the solenoid increases beyond the rating of the circuit breaker, the solenoid's pull releases the latch which then allows the contacts to open by spring action. Some types of magnetic breakers incorporate a hydraulic time delay feature using a viscous fluid.

The core is restrained by a spring until the current exceeds the breaker rating. During an overload, the speed of the solenoid motion is restricted by the fluid. The delay permits brief current surges beyond normal running current for motor starting, energising equipment, etc. Short circuit currents provide sufficient solenoid force to release the latch regardless of core position thus bypassing the delay feature. Ambient temperature affects the time delay but does not affect the current rating of a magnetic breaker

Thermal Magnetic Circuit Breakers

Thermal magnetic circuit breakers, which are the type found in most distribution boards, incorporate both techniques with the electromagnet responding instantaneously to large surges in current (short circuits) and the bimetallic strip responding to less extreme but longer-term over-current conditions. The thermal portion of the circuit breaker provides an "inverse time" response feature which provides faster or slower response for larger or smaller over currents respectively.

Common Trip Breakers

Figure : Three pole common trip breaker for supplying a three-phase device. This breaker has a 2 A rating

When supplying a branch circuit with more than one live conductor, each live conductor must be protected by a breaker pole. To ensure that all live conductors are interrupted when any pole trips, a "common trip" breaker must be used. These may either contain two or three tripping mechanisms within one case, or for small breakers, may externally tie the poles together via their operating handles. Two pole common trip breakers are common on 120/240 volt systems where 240 volt loads (including major appliances or further distribution boards) span the two live wires. Three-pole common trip breakers are typically used to supply three-phase electric power to large motors or further distribution boards.

Two and four pole breakers are used when there is a need to disconnect the neutral wire, to be sure that no current can flow back through the neutral wire from other loads connected to the same network when people need to touch the wires for maintenance.

Separate circuit breakers must never be used for disconnecting live and neutral, because if the neutral gets disconnected while the live conductor stays connected, a dangerous condition arises: the circuit will appear de-energised (appliances will not work), but wires will stay live and RCDs will not trip if someone touches the live wire (because RCDs need power to trip). This is why only common trip breakers must be used when switching of the neutral wire is needed

Medium-voltage Circuit Breakers

Medium-voltage circuit breakers rated between 1 and 72 kV may be assembled into metal-enclosed switchgear line ups for indoor use, or may be individual components installed outdoors in a substation. Air-break circuit breakers replaced oil-filled units for indoor applications, but are now themselves being replaced by vacuum circuit breakers (up to about 35 kV). Like the high voltage circuit breakers described below, these are also operated by current sensing protective relays operated through current transformers.

The characteristics of MV breakers are given by international standards such as IEC 62271. Medium-voltage circuit breakers nearly always use separate current sensors and protective relays, instead of relying on built-in thermal or magnetic overcurrent sensors.

Medium-voltage circuit breakers can be classified by the medium used to extinguish the arc:

- Vacuum circuit breakers—With rated current up to 3000 A, these breakers interrupt the current by creating and extinguishing the arc in a vacuum container. These are generally applied for voltages up to about 35,000 V, which corresponds roughly to the medium-voltage range of power systems. Vacuum circuit breakers tend to have longer life expectancies between overhaul than do air circuit breakers.
- Air circuit breakers—Rated current up to 10,000 A. Trip characteristics are often fully adjustable including configurable trip thresholds and delays. Usually electronically controlled, though some models are microprocessor controlled via an integral electronic trip unit. Often used for main power distribution in large industrial plant, where the breakers are arranged in draw-out enclosures for ease of maintenance.

- SF_6 circuit breakers extinguish the arc in a chamber filled with sulfur hexafluoride gas.

Medium-voltage circuit breakers may be connected into the circuit by bolted connections to bus bars or wires, especially in outdoor switchyards.

Medium-voltage circuit breakers in switchgear line-ups are often built with draw-out construction, allowing the breaker to be removed without disturbing the power circuit connections, using a motor-operated or hand-cranked mechanism to separate the breaker from its enclosure.

High-voltage Circuit Breakers

Figure : Russian 110 kV oil circuit breaker

Electrical power transmission networks are protected and controlled by high-voltage breakers. The definition of *high voltage* varies but in power transmission work is usually thought to be 72.5 kV or higher, according to a recent definition by the International Electrotechnical Commission (IEC). High-voltage breakers are nearly always solenoid-operated, with current sensing protective relays operated through current transformers. In substations the protective relay scheme can be complex, protecting equipment and buses from various types of overload or ground/earth fault.

Figure : 115 kV bulk oil circuit breaker

Figure : 400 kV SF_6 live tank circuit breakers

High-voltage breakers are broadly classified by the medium used to extinguish the arc.

- Bulk oil
- Minimum oil
- Air blast

- Vacuum
- SF_6

Some of the manufacturers are ABB, GE (General Electric), Tavrida Electric, Alstom, Mitsubishi Electric, Pennsylvania Breaker, Siemens, Toshiba, Konèar HVS, BHEL, CGL, Square D (Schneider Electric).

Due to environmental and cost concerns over insulating oil spills, most new breakers use SF_6 gas to quench the arc.

Circuit breakers can be classified as *live tank*, where the enclosure that contains the breaking mechanism is at line potential, or *dead tank* with the enclosure at earth potential. High-voltage AC circuit breakers are routinely available with ratings up to 765 kV. 1200kV breakers were launched by Siemens in November 2011.

High-voltage circuit breakers used on transmission systems may be arranged to allow a single pole of a three-phase line to trip, instead of tripping all three poles; for some classes of faults this improves the system stability and availability.

Sulfur Hexafluoride (SF_6) High-voltage Circuit-breakers

A sulfur hexafluoride circuit breaker uses contacts surrounded by sulfur hexafluoride gas to quench the arc. They are most often used for transmission-level voltages and may be incorporated into compact gas-insulated switchgear. In cold climates, supplemental heating or de-rating of the circuit breakers may be required due to liquefaction of the SF_6 gas.

Other Breakers

The following types are described in separate articles.

- Breakers for protections against earth faults too small to trip an over-current device:
    - Residual-current device (RCD, formerly known as a *residual current circuit breaker*) — detects current imbalance, but does not provide over-current protection.
    - Residual current breaker with over-current protection (RCBO) — combines the functions of an RCD and an MCB in one package. In the United States and Canada, panel-mounted devices that combine ground (earth) fault detection and over-current protection are called Ground Fault Interrupter (GFI) breakers; a wall mounted outlet device or separately enclosed plug-in device providing ground fault

detection and interruption only (no overload protection) is called a Ground Fault Circuit Interrupter (GFCI).

- o Earth leakage circuit breaker (ELCB) — This detects earth current directly rather than detecting imbalance. They are no longer seen in new installations for various reasons.

- Autorecloser — A type of circuit breaker which closes again after a delay. These are used on overhead power distribution systems, to prevent short duration faults from causing sustained outages.
- Polyswitch (polyfuse) — A small device commonly described as an automatically resetting fuse rather than a circuit breaker.

Protective Relay

In electrical engineering, a protective relay is a complex electromechanical apparatus, often with more than one coil, designed to calculate operating conditions on an electrical circuit and trip circuit breakers when a fault is detected. Unlike switching type relays with fixed and usually ill-defined operating voltage thresholds and operating times, protective relays have well-established, selectable, time/current (or other operating parameter) curves. Such relays may be elaborate, using arrays of induction disks, shaded-pole magnets, operating and restraint coils, solenoid-type operators, telephone-relay contacts, and phase-shifting networks. Protection relays respond to such conditions as over-current, over-voltage, reverse power flow, over- and under- frequency. Distance relays trip for faults up to a certain distance away from a substation but not beyond that point. An important transmission line or generator unit will have cubicles dedicated to protection, with many individual electromechanical devices.

Figure : Electromechanical protective relays at a hydroelectric generating plant

The various protective functions available on a given relay are denoted by standard ANSI Device Numbers. For example, a relay including function 51 would be a timed overcurrent protective relay.

Design and theory of these protective devices is an important part of the education of an electrical engineer who specialises in power systems. Today these devices are nearly entirely replaced with microprocessor-based digital protective relays (numerical relays) that emulate their electromechanical ancestors with great precision and convenience in application. By combining several functions in one case, numerical relays also save capital cost and maintenance cost over electromechanical relays. However, due to their very long life span, tens of thousands of these "silent sentinels" are still protecting transmission lines and electrical apparatus all over the world.

Operation Principles

Electromechanical protective relays operate by either magnetic attraction, or magnetic induction.

"Armature"-type relays have a pivoted lever supported on a hinge or knife-edge pivot, which carries a moving contact. These relays may work on either alternating or direct current, but for alternating current, a shading coil on the pole is used to maintain contact force throughout the alternating current cycle. Because the air gap between the fixed coil and the moving armature becomes much smaller when the relay has operated, the current required to maintain the relay closed is much smaller than the current to first operate it. The "returning ratio" or "differential" is the measure of how much the current must be reduced to reset the relay.

A variant application of the attraction principle is the plunger-type or solenoid operator. A reed relay is another example of the attraction principle.

"Moving coil" meters use a loop of wire turns in a stationary magnet, similar to a galvanometer but with a contact lever instead of a pointer. These can be made with very high sensitivity. Another type of moving coil suspends the coil from two conductive ligaments, allowing very long travel of the coil.

"Induction" disk meters work by inducing currents in a disk that is free to rotate; the rotary motion of the disk operates a contact. Induction relays require alternating current; if two or more coils are used, they must be at the same frequency otherwise no net operating force is produced.

Protective relays can also be classified by the type of measurement they make. A protective relay may respond to the magnitude of a quantity such as voltage or current. Induction types of relay can respond to the product of two quantities in two field coils, which could for example represent the power in a circuit. Although an electromechanical relay calculating the ratio of two quantities is not practical, the same effect can be obtained by a balance between two operating coils, which can be arranged to effectively give the same result.

Several operating coils can be used to provide "bias" to the relay, allowing the sensitivity of response in one circuit to be controlled by another. Various combinations of "operate torque" and "restraint torque" can be produced in the relay.

By use of a permanent magnetic in the magnetic circuit, a relay can be made to respond differently to current in one direction than in another. Such polarized relays are used on direct-current circuits to detect, for example, reverse current into a generator. These relays can be made bistable, maintaining a contact closed with no coil current and requiring reverse current to reset.

For AC circuits, the principle is extended with a polarising winding connected to a reference voltage source.

Light weight contacts make for sensitive relays that operate quickly, but small contacts can't carry or break heavy currents. Often auxiliary telephone-type armature relays are triggered by the measuring relay.

In a large installation of electromechanical relays, it would be difficult to determine which device originated the signal that tripped the circuit. This information is useful to operating personnel to determine the likely cause of the fault and to prevent its re-occurrence. Relays may be fitted with a "target" or "flag" unit, which is released when the relay operates, to display a distinctive coloured signal when the relay has tripped.

Types

Overcurrent Relay

An "overcurrent relay" is a type of protective relay which operates when the load current exceeds a preset value. The ANSI device number is 50 for an instantaneous overcurrent (IOC), 51 for a time over current (TOC). In a typical application the overcurrent relay is connected to a current transformer and calibrated to operate at or above a specific current level. When the relay operates, one or more contacts will operate and energise to trip (open) a circuit breaker.

Induction Disc Overcurrent Relay

These robust and reliable electromagnetic relays use the induction principle discovered by Galileo Ferraris in the late 19th century. The magnetic system in induction disc overcurrent relays is designed to detect overcurrents in a power system and operate with a pre-determined time delay when certain overcurrent limits have been reached. In order to operate, the magnetic system in the relays produces torque that acts on a metal disc to make contact, according to the following basic current/torque equation:

$$T = K \times \phi_1 \times \phi_2 \sin\theta$$

Where

K– is a constant ϕ_1 and ϕ_2 are the two fluxes θ is the phase angle between the fluxes.

The relay's primary winding is supplied from the power systems current transformer via a plug bridge, which is called the plug setting multiplier (psm). Usually seven equally spaced tappings or operating bands determine the relays sensitivity. The primary winding is located on the upper electromagnet. The secondary winding has connections on the upper electromagnet that are energised from the primary winding and connected to the lower electromagnet. Once the upper and lower electromagnets are energised they produce eddy currents that are induced onto the metal disc and flow through the flux paths. This relationship of eddy currents and fluxes creates torque proportional to the input current of the primary winding, due to the two flux paths been out of phase by 90°.

In an overcurrent condition, a value of current will be reached that overcomes the control spring pressure on the spindle and the braking magnet, causing the metal disc to rotate towards the fixed contact. This initial movement of the disc is also held off to a critical positive value of current by small slots that are often cut into the side of the disc. The time taken for rotation to make the contacts is not only dependent on current but also the spindle backstop position, known as the time multiplier (tm). The time multiplier is divided into 10 linear divisions of the full rotation time.

Providing the relay is free from dirt, the metal disc and the spindle with its contact will reach the fixed contact, thus sending a signal to trip and isolate the circuit, within its designed time and current specifications. Drop off current of the relay is much lower than its operating value, and once reached the relay will be reset in a reverse motion by the pressure of the control spring governed by the braking magnet.

Distance Relay

The most common form of protection on high voltage transmission systems is distance relay protection. Power lines have set impedance per kilometre and using this value and comparing voltage and current the distance to a fault can be determined. The ANSI standard device number for a distance relay is 21.

Current Differential Protection

Another common form of protection for apparatus such as transformers, generators, busses and power lines is current differential. This type of protection works on the basic theory of Kirchhoff's current law which states that the sum of the currents entering and exiting a node will equal zero. It is important to note the direction of the currents as well as the magnitude, as they are vectors. It requires a set of current transformers (smaller transformers that transform currents down to a level which can be measured) at each end of the power line, or each side of the transformer. The current protection relay then compares the currents and calculates the difference between the two. As an example, a power line from one substation to another will have a current differential relay at both substations which communicate with each other. In a healthy condition, the relay at substation A may read 500 amps (power exporting) and substation B will read 500 amps (power importing). If a path to earth or ground develops there will be a surge of current. As supply grids are generally well interconnected the fault in the previous example will be fed from both ends of the power line. The relay at substation A will see a massive increase in current and will continue to export. Substation B will also see a massive increase in current, however it will now start to export as well. In turn the protection relay will see the currents travelling in opposite directions (180 degrees phase shift) and instead of cancelling each other out to give a summation of zero it will see a large value of current. The relays will trip the associated circuit breakers. This type of protection is called unit protection, as it only protects what is between the current transformers. It is important to note that generally the higher the currents in the lines the larger the differential current required for the relay to see it as a fault. This is basically done due to small mismatches in current transformers.

Small errors will increase as current increases to the point where the error could cause a false trip if the current differential relay only had an upper limit instead of the rising differential characteristic. It is also important to note that CTs have a point where the core saturates and the current in the CT is no longer proportional to the current in the line. A CT can become inaccurate or even saturate because of a fault outside of its protected zone (through fault) where the CTs see a large magnitude but still in the same direction.

Static Relays

Application of electronic amplifiers to protective relays was described as early as 1928, using vacuum tube amplifiers. Devices using electron tubes were studied but never applied as commercial products, because of the limitations of vacuum tube amplifiers. A relatively large standby current is required to maintain the tube filament temperature; inconvenient high voltages are required for the circuits, and vacuum tube amplifiers had difficulty with incorrect operation due to noise disturbances. Static relays with no or few moving parts became practical with the introduction of the transistor. Static relays offer the advantage of higher sensitivity than purely electromechanical relays, because power to operate output contacts is derived from a separate supply, not from the signal circuits. Static relays eliminated or reduced contact bounce, and could provide fast operation, long life and low maintenance.

Digital Protective Relays

The functions of electromechanical protection systems are now being replaced by microprocessor-based digital protective relays, sometimes called "numeric relays".

Protective Devices

- Protective relays control the tripping of the circuit breakers surrounding the faulted part of the network
- Automatic operation, such as auto-reclosing or system restart
- Monitoring equipment which collects data on the system for post event analysis

While the operating quality of these devices, and especially of protective relays, is always critical, different strategies are considered for protecting the different parts of the system. Very important equipment may have completely redundant and independent protective systems, while a minor branch distribution line may have very simple low-cost protection.

Types of Protection

- Generator sets – In a power plant, the protective relays are intended to prevent damage to alternators or to the transformers in case of abnormal conditions of operation, due to internal failures, as well as insulating failures or regulation malfunctions. Such failures are unusual, so the protective relays have to operate very rarely. If a protective relay fails to detect a fault, the resulting damage to the alternator or to the transformer might require costly equipment repairs or replacement, as well as income loss from the inability to produce and sell energy.
- High voltage transmission network – Protection on the transmission and distribution serves two functions: Protection of plant and protection of the public (including employees). At a basic level, protection looks to disconnect equipment which experience an overload or a short to earth. Some items in substations such as transformers might require additional protection based on temperature or gas pressure, among others.
- Overload & Backup for Distance (Overcurrent) – Overload protection requires a current transformer which simply measures the current in a circuit. There are two types of overload protection: instantaneous overcurrent and time overcurrent (TOC). Instantaneous overcurrent requires that the current exceeds a pre-determined level for the circuit breaker to operate. TOC protection operates based on a current vs time curve. Based on this curve if the measured current exceeds a given level for the preset amount of time, the circuit breaker or fuse will operate.
- Earth fault – Earth fault protection again requires current transformers and senses an imbalance in a three-phase circuit. Normally the three phase currents are in balance, i.e. roughly equal in magnitude. If one or two phases become connected to earth via a low impedance path, their magnitudes will increase dramatically, as will current imbalance. If this imbalance exceeds a pre-determined value, a circuit breaker should operate.
- Distance (Impedance Relay)– Distance protection detects both voltage and current. A fault on a circuit will generally create a sag in the voltage level. If the ratio of voltage to current measured at the relay terminals, which equates to an

impedance, lands within a pre-determined level the circuit breaker will operate. This is useful for reasonable length lines, lines longer than 10 miles, because its operating characteristics are based on the line characteristics. This means that when a fault appears on the line the impedance setting in the relay is compared to the apparent impedance of the line from the relay terminals to the fault. If the relay setting is determined to be below the apparent impedance it is determined that the fault is within the zone of protection. When the transmission line length is too short, less than 10 miles, distance protection is becomes more difficult to coordinate.

- Back-up – At all times the objective of protection is to remove only the affected portion of plant and nothing else. Sometimes this does not occur for various reasons which can include:
  - o Mechanical failure of a circuit breaker to operate
  - o Incorrect protection setting
  - o Relay failures

A failure of primary protection will usually result in the operation of back-up protection. Remote back-up protection will generally remove both the affected and unaffected items of plant to clear the fault. Local back-up protection will remove the affected items of the plant to clear the fault.

- Low-voltage networks – The low voltage network generally relies upon fuses or low-voltage circuit breakers to remove both overload and earth faults.

Coordination

Protective device coordination is the process of determining the "best fit" timing of current interruption when abnormal electrical conditions occur. The goal is to minimise an outage to the greatest extent possible. Historically, protective device coordination was done on translucent log-log paper. Modern methods normally include detailed computer based analysis and reporting.

Disturbance Monitoring Equipment

Disturbance monitoring equipment (DME) monitors and records system data pertaining to a fault. DME accomplish three main purposes:

1. model validation,
2. disturbance investigation, and
3. assessment of system protection performance.

DME devices include:

- Sequence of event recorders, which record equipment response to the event
- Fault recorders, which record actual waveform data of the system primary voltages and currents.
- Dynamic Disturbance Recorders (DDRs), which record incidents that portray power system behaviour during dynamic events such as low frequency (0.1 Hz – 3 Hz) oscillations and abnormal frequency or voltage excursions

Performance Measures

Protection engineers define dependability as the tendency of the protection system to operate correctly for in-zone faults. They define security as the tendency not to operate for out-of-zone faults. Both dependability and security are reliability issues. Fault tree analysis is one tool with which a protection engineer can compare the relative reliability of proposed protection schemes.

Quantifying protection reliability is important for making the best decisions on improving a protection system, managing dependability versus security tradeoffs, and getting the best results for the least money. A quantitative understanding is essential in the competitive utility industry.

Slack Bus

A 'Slack' bus (or 'swing' bus) is defined as Vä bus, that is used to balance the active |P| & reactive power |Q| in the system while performing Load flow studies in Electrical Power Systems. For Power Engineers, Load flow study is very important as it explains the power system conditions at various intervals during operation.

Slack Bus is used to provide system losses by emitting or absorbing active/reactive power to/from the system. While this definition of the load flow problem is appropriate for a deterministic solution, it has an inherent drawback when dealing with uncertain input variables: the slack bus must absorb all uncertainties arising from the system and thus, will have the widest nodal power possibility (probability) distributions in the system.

If even moderate amounts of uncertainty are allowed in a large system, the resulting distributions will frequently contain values well beyond the generating margins of the slack generator.

Description

The most common formulation of the Load Flow problem requires all input variables (PQ at loads, PV at generators) to be specified as deterministic values. Each set of specified values corresponds to one system state, which depends on some set of system conditions. Thus, when the input conditions are uncertain, there is a need for numerous scenarios to be analysed. A load flow approach that could directly incorporate uncertainty into the solution process has been long recognised as useful. The results from such analysis would be expected to give solutions over the range of the uncertainties, i.e., solutions that are sets of values or regions instead of single operating points.

Load buses are of 3 types and are classified as :

1. PQ buses - here, the real power $|P|$ and reactive power $|Q|$ are specified. It is also known as *Load Bus*.
2. PV buses - here, the real power $|P|$ and the voltage magnitude $|V|$ are specified. It is also known as *Generator Bus*.
3. Slack bus - to balance the active and reactive power in the system. It is also known as the *Reference Bus* or the *Swing Bus*.

The Slack bus is selected to provide/absorb active and reactive power to/from the transmission line to provide losses since these are unknown till the final solution of the problem. The slack bus is the only bus at which the system reference phase angle is defined. From this, the various angular differences can be calculated in the power flow equations. If a slack bus is not specified, then a generator bus with maximum real power $|P|$ will be chosen as the Slack bus. There can be more than one slack bus in a given scheme.

Formulation of Load Flow Problem

A classic load flow problem consists of calculation of voltage magnitude and its phase angle at the buses, and also the active and reactive line flows for the specified terminal or bus conditions. Associated with each bus of a power system, there are 4 set of variables:

1. magnitude of voltage i.e. $|V|$
2. phase angle i.e. $|\delta|$
3. active or real power i.e. $|P|$
4. reactive power i.e. $|Q|$

Based on these values, a bus may be classified into the above mentioned three categories as -

	P	Q	V	δ
P-Q bus	known	known	unknown	unknown
P-V bus	known	unknown	known	unknown
Slack bus	unknown	unknown	known	known

The solution of load flow problem requires two main steps viz 'mathematical formulation of the problem' and 'application of numerical technique to solve the problem'. In a load flow study real and reactive powers (i.e. complex power) cannot be fixed at all the buses as the net complex power flow into the network is not known in advance, the system power loss being unknown till the load flow study is complete. It is, therefore, necessary to have one bus (i.e. the slack bus) at which complex power is unspecified so that it supplies the difference in the total system load plus losses and the sum of the complex powers specified at the remaining buses. By the same reasoning the slack bus must be a generator bus. The complex power allocated to this bus is determined as part of the solution. In order that the variations in real and reactive powers of the slack bus during the iterative process be a small percentage of its generating capacity, the bus connected to the largest generating station is normally selected as the slack bus.

The load flow problems generate non-linear equations but as we know computers cant solve these equations directly, there is a need for a particular numerical method. To solve the non-linear equations, we have three methods -

1. Gauss Iterative Method.
2. Fast Decoupled Load Flow Method.
3. Gauss-Seidel Method.
4. Newton-Raphson Method.

Power Quality Compression Algorithm

A Power Quality Compression Algorithm is an algorithm used in power quality analysis. Electricity today is classed as one of the most important commodities, and has a particular characteristic that it is either impossible or uneconomical to store. Hence, electricity must be consumed as soon as it is produced. In order to provide high quality electric power service, it is essential to monitor the quality of the electric signals also termed as Power Quality (PQ) at different locations along an electrical power network. Electrical utilities carefully monitor waveforms and currents at various network locations constantly, in order for them to understand what lead up to any unforeseen events such as a power outage and blackouts. This is particularly critical at sites where the environment and public safety are at risk (institutions such as hospitals, sewage treatment plants, mines, etc.).

Power Quality Challenges

Engineers have at their disposal many meters, that are able to read and display electrical power waveforms and calculating parameters of the waveforms. These parameters may include, for example, current and voltage RMS, phase relationship between waveforms of a multi-phase signal, power factor, frequency, power factor, THD, harmonics, active power (KWatt), reactive power (KVAR), apparent power (KVA) and active energy (KWh), reactive energy (KVARH) and apparent energy (KVAh) and many more. In order to sufficiently monitor unforeseen events, Ribeiro et al. explains that it is not enough to display these parameters, but to also capture voltage waveform data at all times. This virtually impracticable due to the large amounts of data involved, causing what is known the "bottle effect". For instance at a sampling rate of 32 samples per cycle, 1,920 samples are collected per second. For three-phase meters that measure both voltage and current waveforms, the data is 6-8 times as much. More practical solutions developed in recent years, store data only when an event occurs (e.g., High levels of power system harmonics that creates voltage distortion) or alternatively to store the RMS of power the electrical signals. This data, however, is not always sufficient to determine the exact nature of problems.

Data Compression Algorithm

Nisenblat *et al.* proposes the idea of Power Quality (PQ) compression algorithm (similar to the Lossy Compression method) that enables meters to continuously store the waveform of one or more power signals, regardless whether or not an event of interest was identified. This algorithm referred to as PQZip empowers a processor with a memory that is sufficient enough to store the waveform, under normal power conditions, over a long period of time, of at least a month, two months or even a year. The compression is performed in real time, as the signals are acquired; it calculates a compression decision before all the compressed data is received. For instance should 1 parameter remain constant, and various others fluctuate, the compression decision retains only what is relevant from the constant data, and retains all the fluctuation data. It then decomposes the waveform of the power signal of numerous components, over various periods of the waveform. It concludes the process by compressing the values of at least some of these components over different periods, separately. This real time compression algorithm, performed independent of the sampling, prevents data gaps and has a typical 1000:1 compression ratio.

Conclusion-Importance of Compression Algorithm

Optimising the distribution and consumption of our energy resources in particular electricity, is vital for the preservation of such resources, especially considering the status of the planet today. Development aforementioned ideas have been crucial in understanding and prevention of unforeseen events during the distribution and consumption of electricity. The challenge today is not only to identify that an event happened and when the event happened. It is also crucial to have sufficient data at hand to ensure that contingency plans are in place, and preventative action will be taken. Efficient compression algorithm prevents "bottle effect" mentioned above, and ensures that data is considered and processed responsibly for efficient and effective use.

Linear Programming

Linear programming (LP, or linear optimisation) is a mathematical method for determining a way to achieve the best outcome (such as maximum profit or lowest cost) in a given mathematical model for some list of requirements represented as linear relationships. Linear programming is a specific case of mathematical programming (mathematical optimisation).

More formally, linear programming is a technique for the optimisation of a linear objective function, subject to linear equality and linear inequality constraints. Its feasible region is a convex polyhedron, which is a set defined as the intersection of finitely many half spaces, each of which is defined by a linear inequality. Its objective function is a real-valued affine function defined on this polyhedron. A linear programming algorithm finds a point in the polyhedron where this function has the smallest (or largest) value if such point exists.

Linear programs are problems that can be expressed in canonical form:

$$\begin{aligned} &\text{maximize} && \mathbf{c}^{\mathrm{T}}\mathbf{x} \\ &\text{subject to} && A\mathbf{x} \leq \mathbf{b} \\ &\text{and} && \mathbf{x} \geq \mathbf{0} \end{aligned}$$

where x represents the vector of variables (to be determined), c and b are vectors of (known) coefficients and A is a (known) matrix of coefficients. The expression to be maximised or minimised is called the *objective function* ($c^T x$ in this case). The equations $Ax \leq b$ are the constraints which specify a convex polytope over which the objective function is to be optimised. (In this context, two vectors are comparable when every entry in one is less-than or equal-to the corresponding entry in the other. Otherwise, they are incomparable.)

Linear programming can be applied to various fields of study. It is used in business and economics, but can also be utilised for some engineering problems. Industries that use linear programming models include transportation, energy, telecommunications, and manufacturing. It has proved useful in modelling diverse types of problems in planning, routing, scheduling, assignment, and design.

History

The problem of solving a system of linear inequalities dates back at least as far as Fourier, after whom the method of Fourier-Motzkin elimination is named. The earliest linear programming was first developed by Leonid Kantorovich in 1939. Leonid Kantorovich developed the earliest linear programming problems in 1939 for use during World War II to plan expenditures and returns in order to reduce costs to the army and increase losses to the enemy. The method was kept secret until 1947 when George B. Dantzig published the simplex method and John von Neumann developed the theory of duality as a linear optimisation solution, and applied it in the field of game theory. Postwar, many industries found its use in their daily planning.

The linear-programming problem was first shown to be solvable in polynomial time by Leonid Khachiyan in 1979, but a larger theoretical and practical breakthrough in the field came in 1984 when Narendra Karmarkar introduced a new interior-point method for solving linear-programming problems.

Dantzig's original example was to find the best assignment of 70 people to 70 jobs. The computing power required to test all the permutations to select the best assignment is vast; the number of possible configurations exceeds the number of particles in the universe. However, it takes only a moment to find the optimum solution by posing the problem as a linear program and applying the Simplex algorithm. The theory behind linear programming drastically reduces the number of possible optimal solutions that must be checked.

Uses

Linear programming is a considerable field of optimisation for several reasons. Many practical problems in operations research can be expressed as linear programming problems. Certain special cases of linear programming, such as *network flow* problems and *multicommodity flow* problems are considered important enough to have generated much research on specialised algorithms for their solution.

A number of algorithms for other types of optimisation problems work by solving LP problems as sub-problems. Historically, ideas from linear programming have inspired many of the central concepts of optimisation theory, such as *duality, decomposition,* and the importance of *convexity* and its generalisations. Likewise, linear programming is heavily used in microeconomics and company management, such as planning, production, transportation, technology and other issues. Although the modern management issues are ever-changing, most companies would like to maximise profits or minimise costs with limited resources. Therefore, many issues can be characterised as linear programming problems.

Standard form

Standard form is the usual and most intuitive form of describing a linear programming problem. It consists of the following four parts:

- A linear function to be maximised

  e.g. $f(x_1, x_2) = c_1x_1 + c_2x_2$

- Problem constraints of the following form

  e.g. $$\begin{aligned} a_{11}x_1 + a_{12}x_2 &\le b_1 \\ a_{21}x_1 + a_{22}x_2 &\le b_2 \\ a_{31}x_1 + a_{32}x_2 &\le b_3 \end{aligned}$$

- Non-negative variables

  e.g. $$\begin{aligned} x_1 &\ge 0 \\ x_2 &\ge 0 \end{aligned}$$

- Non-negative right hand side constants

  $b_i \ge 0, i = 1, 2, 3$

The problem is usually expressed in *matrix form*, and then becomes:

$$\max\{c^{\mathrm{T}}x \mid 0 \le Ax \le b \land x \ge 0\}$$

Other forms, such as minimisation problems, problems with constraints on alternative forms, as well as problems involving negative variables can always be rewritten into an equivalent problem in standard form.

Theory

Existence of Optimal Solutions

Geometrically, the linear constraints define the feasible region, which is a convex polyhedron. A linear function is a convex function, which implies that every local minimum is a global minimum; similarly, a linear function is a concave function, which implies that every local maximum is a global maximum.

Optimal solution need not exist, for two reasons. First, if two constraints are inconsistent, then no feasible solution exists: For instance, the constraints $x \geq 2$ and $x \leq 1$ cannot be satisfied jointly; in this case, we say that the LP is *infeasible*. Second, when the polytope is unbounded in the direction of the gradient of the objective function (where the gradient of the objective function is the vector of the coefficients of the objective function), then no optimal value is attained.

Optimal Vertices (and Rays) of Polyhedra

Otherwise, if a feasible solution exists and if the (linear) objective function is bounded, then the optimum value is always attained on the boundary of optimal level-set, by the *maximum principle* for *convex functions* (alternatively, by the *minimum* principle for *concave functions*): Recall that linear functions are both convex and concave. However, some problems have distinct optimal solutions: For example, the problem of finding a feasible solution to a system of linear inequalities is a linear programming problem in which the objective function is the zero function (that is, the constant function taking the value zero everywhere): For this feasibility problem with the zero-function for its objective-function, if there are two distinct solutions, then every convex combination of the solutions is a solution.

The vertices of the polytope are also called *basic feasible solutions*. The reason for this choice of name is as follows. Let d denote the number of variables. Then the fundamental theorem of linear inequalities implies (for feasible problems) that for every vertex x^* of the LP feasible region, there exists a set of d (or fewer) inequality constraints from the LP such that, when we treat those d constraints as equalities, the unique solution is x^*. Thereby we can study these vertices by means of looking at certain subsets of the set of all constraints (a discrete set), rather than the continuum of LP solutions. This principle underlies the simplex algorithm for solving linear programs.

Algorithms

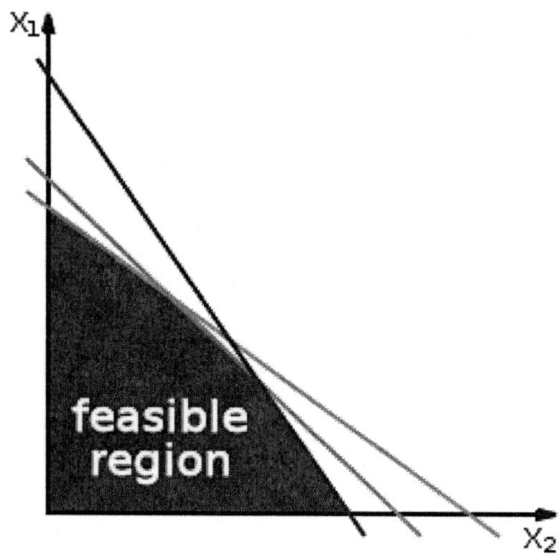

Figure : A series of linear constraints on two variables produces a region of possible values for those variables. Solvable problems will have a feasible region in the shape of a simple polygon.

Basis Exchange Algorithms

Simplex Algorithm of Dantzig

The simplex algorithm, developed by George Dantzig in 1947, solves LP problems by constructing a feasible solution at a vertex of the polytope and then walking along a path on the edges of the polytope to vertices with non-decreasing values of the objective function until an optimum is reached. In many practical problems, "stalling" occurs: Many pivots are made with no increase in the objective function. In rare practical problems, the usual versions of the simplex algorithm may actually "cycle". To avoid cycles, researchers developed new pivoting rules.

In practice, the simplex algorithm is quite efficient and can be guaranteed to find the global optimum if certain precautions against *cycling* are taken. The simplex algorithm has been proved to solve "random" problems efficiently, i.e. in a cubic number of steps, which is similar to its behaviour on practical problems.

However, the simplex algorithm has poor worst-case behaviour: Klee and Minty constructed a family of linear programming problems for which the simplex method takes a number of steps exponential in the problem size. In fact, for some time it was not known whether the linear programming problem was solvable in polynomial time, i.e. of complexity class P.

Criss-cross Algorithm

Like the simplex algorithm of Dantzig, the criss-cross algorithm is a basis-exchange algorithm that pivots between bases. However, the criss-cross algorithm need not maintain feasibility, but can pivot rather from a feasible basis to an infeasible basis. The criss-cross algorithm does not have polynomial time-complexity for linear programming. Both algorithms visit all 2^D corners of a (perturbed) cube in dimension D, the Klee–Minty cube (after Victor Klee and George J. Minty), in the worst case.

Interior Point

Ellipsoid Algorithm, Following Khachiyan

This is the first worst-case polynomial-time algorithm for linear programming. To solve a problem which has n variables and can be encoded in L input bits, this algorithm uses $O(n^4L)$ pseudo-arithmetic operations on numbers with $O(L)$ digits. Khachiyan's algorithm and his long standing issue was resolved by Leonid Khachiyan in 1979 with the introduction of the ellipsoid method. The convergence analysis have (real-number) predecessors, notably the iterative methods developed by Naum Z. Shor and the approximation algorithms by Arkadi Nemirovski and D. Yudin.

Projective Algorithm of Karmarkar

Khachiyan's algorithm was of landmark importance for establishing the polynomial-time solvability of linear programs. The algorithm was not a computational break-through, as the simplex method is more efficient for all but specially constructed families of linear programs.

However, Khachiyan's algorithm inspired new lines of research in linear programming. In 1984, N. Karmarkar proposed a projective method for linear programming. Karmarkar's algorithm improved on Khachiyan's worst-case polynomial bound (giving $O(n^{3.5}L)$). Karmarkar claimed that his algorithm was much faster in practical LP than the simplex method, a claim that created great interest in interior-point methods. Its projective geometry is interesting.

Path-following Algorithms

In contrast to the simplex algorithm, which finds an optimal solution by traversing the edges between vertices on a polyhedral set, interior-point methods move through the interior of the feasible region. Since then, many interior-point methods have been proposed and analysed. Early successful implementations were based on *affine scaling* variants of the method. For both theoretical and practical purposes, barrier function or path-following methods have been the most popular since the 1990s.

Comparison of Interior-point Methods versus Simplex Algorithms

The current opinion is that the efficiency of good implementations of simplex-based methods and interior point methods are similar for routine applications of linear programming. However, for specific types of LP problems, it may be that one type of solver is better than another (sometimes much better).

LP solvers are in widespread use for optimisation of various problems in industry, such as optimisation of flow in transportation networks.

Open Problems and Recent Work

There are several open problems in the theory of linear programming, the solution of which would represent fundamental breakthroughs in mathematics and potentially major advances in our ability to solve large-scale linear programs.

- Does LP admit a strongly polynomial-time algorithm?
- Does LP admit a strongly polynomial algorithm to find a strictly complementary solution?
- Does LP admit a polynomial algorithm in the real number (unit cost) model of computation?

This closely related set of problems has been cited by Stephen Smale as among the 18 greatest unsolved problems of the 21st century. In Smale's words, the third version of the problem "is the main unsolved problem of linear programming theory." While algorithms exist to solve linear programming in weakly polynomial time, such as the ellipsoid methods and interior-point techniques, no algorithms have yet been found that allow strongly polynomial-time performance in the number of constraints and the number of variables. The development of such algorithms would be of great theoretical interest, and perhaps allow practical gains in solving large LPs as well.

Although the Hirsch conjecture was recently disproved for higher dimensions, it still leaves the following questions open.

- Are there pivot rules which lead to polynomial-time Simplex variants?
- Do all polytopal graphs have polynomially-bounded diameter?

These questions relate to the performance analysis and development of Simplex-like methods. The immense efficiency of the Simplex algorithm in practice despite its exponential-time theoretical performance hints that there may be variations of Simplex that run in polynomial or even strongly polynomial time. It would be of great practical and theoretical significance to know whether any such variants exist, particularly as an approach to deciding if LP can be solved in strongly polynomial time.

The Simplex algorithm and its variants fall in the family of edge-following algorithms, so named because they solve linear programming problems by moving from vertex to vertex along edges of a polytope. This means that their theoretical performance is limited by the maximum number of edges between any two vertices on the LP polytope. As a result, we are interested in knowing the maximum graph-theoretical diameter of polytopal graphs. It has been proved that all polytopes have subexponential diameter. The recent disproof of the Hirsch conjecture is the first step to prove whether any polytope has superpolynomial diameter. If any such polytopes exist, then no edge-following variant can run in polynomial time. Questions about polytope diameter are of independent mathematical interest.

Simplex pivot methods preserve primal (or dual) feasibility. On the other hand, criss-cross pivot methods do not preserve (primal or dual) feasibility—they may visit primal feasible, dual feasible or primal-and-dual infeasible bases in any order. Pivot methods of this type have been studied since the 1970s. Essentially, these methods attempt to find the shortest pivot path on the arrangement polytope under the linear programming problem. In contrast to polytopal graphs, graphs of arrangement polytopes are known to have small diameter, allowing the possibility of strongly polynomial-time criss-cross pivot algorithm without resolving questions about the diameter of general polytopes.

Integer Unknowns

If the unknown variables are all required to be integers, then the problem is called an integer programming (IP) or integer linear programming (ILP) problem. In contrast to linear programming, which can be solved efficiently in the worst case, integer programming problems are in many practical situations (those with bounded variables) NP-hard. 0-1 integer programming or binary integer programming (BIP) is the special case of integer programming where variables are required to be 0 or 1 (rather than arbitrary integers). This problem is also classified as NP-hard, and in fact the decision version was one of Karp's 21 NP-complete problems.

If only some of the unknown variables are required to be integers, then the problem is called a mixed integer programming (MIP) problem. These are generally also NP-hard.

There are however some important subclasses of IP and MIP problems that are efficiently solvable, most notably problems where the constraint matrix is totally unimodular and the right-hand sides of the constraints are integers.

Advanced algorithms for solving integer linear programs include:

- cutting-plane method
- branch and bound
- branch and cut
- branch and price
- if the problem has some extra structure, it may be possible to apply delayed column generation.

Such integer-programming algorithms are discussed by Padberg and in Beasley.

Integral Linear Programs

A linear program in real variables is said to be integral if it has at least one optimal solution which is integral. Likewise, a polyhedron $P = \{x \mid Ax \geq 0\}$ is said to be integral if for all bounded feasible objective functions c, the linear program $\{\max cx \mid x \in P\}$ has an optimum x^* with integer coordinates. As observed by Edmonds and Giles in 1977, one can equivalently say that a polyhedron is integral if for every bounded feasible integral objective function c, the optimal *value* of the linear program $\{\max cx \mid x \in P\}$ is an integer.

Integral linear programs are of central importance in the polyhedral aspect of combinatorial optimisation since they provide an alternate characterisation of a problem. Specifically, for any problem, the convex hull of the solutions is an integral polyhedron; if this polyhedron has a nice/compact description, then we can efficiently find the optimal feasible solution under any linear objective. Conversely, if we can prove that a linear programming relaxation is integral, then it is the desired description of the convex hull of feasible (integral) solutions.

Note that terminology is not consistent throughout the literature, so one should be careful to distinguish the following two concepts,

- in an *integer linear program,* described in the previous section, variables are forcibly constrained to be integers, and this problem is NP-hard in general,
- in an *integral linear program,* described in this section, variables are not constrained to be integers but rather one has proven somehow that the continuous problem always has an integral optimal value (assuming c is integral), and this optimal value may be found efficiently since all polynomial-size linear programs can be solved in polynomial time.

One common way of proving that a polyhedron is integral is to show that it is totally unimodular. There are other general methods including the integer decomposition property and total dual integrality. Other specific well-known integral LPs include the matching polytope, lattice polyhedra, submodular flow polyhedra, and the intersection of 2 generalised polymatroids/g-polymatroids.

A bounded integral polyhedron is sometimes called a convex lattice polytope, particularly in two dimensions.

Solvers and Scripting (Programming) Languages

Free open-source permissive licenses:

Name	*License*	*Brief info*
OpenOpt	BSD	Universal cross-platform numerical optimisation framework
pulp-or	BSD	Python module for modelling and solving linear programming problems
Pyomo	BSD	Python module for formulating linear programming problems with abstract models
JOptimiser	Apache V2.0	Java library for convex optimisation

Free open-source copyleft (reciprocal) licenses:

Name	*License*	*Brief info*
LP_Solve	LGPL	User-friendly linear and integer programming solver. Also provides DLL for program integration, and is compatible with GNU MathProg and Zimpl modelling languages.

Cassowary constraint solver	LGPL	an incremental constraint solving toolkit that efficiently solves systems of linear equalities and inequalities.
CVXOPT	GPL	general purpose convex optimisation solver written in Python, with a C API, and calls external routines (e.g. BLAS, LAPACK, FFTW) for numerical computations. Has its own solvers, but can also call glpk or MOSEK if installed
glpk	GPL	GNU Linear Programming Kit, a free LP/MILP solver. Uses GNU MathProg modelling language.
Qoca	GPL	a library for incrementally solving systems of linear equations with various goal functions
CBC	CPL	a MIP solver from COIN-OR
CLP	CPL	an LP solver from COIN-OR
R-Project	GPL	a programming language and software environment for statistical computing and graphics
CVX	GPL	MATLAB based modelling system for convex optimisation, including linear programs; calls either SDPT3 or SeDuMi as a solver
CVXMOD	GPL	Python based modelling system, similar to CVX. It calls CVXOPT as its solver. It is still in alpha release, as of 2009
SDPT3	GPL	MATLAB based convex optimisation solver
SeDuMi	GPL	MATLAB based convex optimisation solver

MINTO (Mixed Integer Optimiser, an integer programming solver which uses branch and bound algorithm) has publicly available source code but not open source.

Proprietary:

Name	*Brief Info*
APMonitor	
AIMMS	
AMPL	A popular modelling language for large-scale linear, mixed integer and nonlinear optimisation with a free student version available.
Analytica	Optimisation modelling software that incorporates state-of-the-art algorithms for

	linear and nonlinear optimisation. Supports LP, NLP, QP, continuous and integer optimisation.
CPLEX	Popular solver with an API for several programming languages, and also has a modelling language and works with AIMMS, AMPL, GAMS, MPL, OpenOpt, OPL Development Studio, and TOMLAB
EXCEL Solver Function	
FinMath	A .NET numerical library containing an interior-point primal-dual solver.
FortMP	
GAMS	
GIPALS	The maximum number of constraints and variables is unlimited. The linear program solver is based on Interior-Point method (Mehrotra predictor-corrector algorithm) and optimised for large sparse linear programs by implementing the state-of-art algorithm to order the constraints matrix. The user can specify the linear program using a set of exported DLL functions.
Gurobi	Solver with parallel algorithms for large-scale linear programs, quadratic programs and mixed-integer programs. Free for academic use.
IMSL Numerical Libraries	Collections of math and statistical algorithms available in C/C++, Fortran, Java and C#/.NET. Optimisation routines in the IMSL Libraries include unconstrained, linearly and nonlinearly constrained minimisations, and linear programming algorithms.
Lingo	
LPL	
LiPS (freeware)	Linear Program Solver (LiPS) is intended for solving linear programming problems. Main features: easy to use graphical interface, sensitivity analysis, goal and mixed integer programming solver. LiPS supports MPS and simple LP format (like lpsolve).
MATLAB	A general-purpose and matrix-oriented programming-language for numerical computing. Linear programming in MATLAB requires the Optimisation Toolbox in addition to the base MATLAB product; available routines include BINTPROG and LINPROG

Mathematica	A general-purpose programming-language for mathematics, including symbolic and numerical capabilities.
MOPS MOSEK	A solver for large scale optimisation with API for several languages (C++, java, .net, Matlab and python).
NAG Numerical Library	A collection of mathematical and statistical routines developed by the Numerical Algorithms Group for multiple programming languages (C, C++, Fortran, Visual Basic, Java and C#) and packages (MATLAB, Excel, R, LabVIEW). The NAG Library includes routines for linear programming problems with both sparse and non-sparse linear constraint matrices, together with routines for the optimisation of quadratic, nonlinear, sums of squares of linear or nonlinear functions with nonlinear, bounded or no constraints. The NAG Library has routines for both local and global optimisation, and for continuous or integer problems.
NMath Stats	A general-purpose .NET statistical library containing a simplex solver.
OptimJ	A Java-based modelling language for optimisation with a free version available.
SAS/OR	A suite of solvers for Linear, Integer, Nonlinear, Derivative-Free, Network, Combinatorial and Constraint Optimisation; the Algebraic modelling language OPTMODEL; and a variety of vertical solutions aimed at specific problems/markets, all of which are fully integrated with the SAS System.
SCIP	A general-purpose constraint integer programming solver with an emphasis on MIP. Compatible with Zimpl modelling language. Free for academic use and available in source code.
Solver Foundation	A .NET platform for modelling, scheduling, and optimisation.
SoPlex	The Sequential object-oriented simPlex: a general-purpose LP solver. Free for academic use and available in source code.
SuanShu	A Java-based math library that supports linear programming and other kinds of numerical optimisation.

TOMLAB	
VisSim	A visual block diagram language for simulation of dynamical systems.

Nonlinear Programming

In mathematics, nonlinear programming (NLP) is the process of solving a system of equalities and inequalities, collectively termed constraints, over a set of unknown real variables, along with an objective function to be maximised or minimised, where some of the constraints or the objective function are nonlinear.

Applicability

A typical nonconvex problem is that of optimising transportation costs by selection from a set of transportion methods, one or more of which exhibit economies of scale, with various connectivities and capacity constraints.

An example would be petroleum product transport given a selection or combination of pipeline, rail tanker, road tanker, river barge, or coastal tankship.

Owing to economic batch size the cost functions may have discontinuities in addition to smooth changes.

Mathematical Formulation of the Problem

The problem can be stated simply as:

$\max_{x \in X} f(x)$ to maximise some variable such as product throughput

or

$\max_{x \in X} f(x)$ to minimise a cost function

where

$$f : R^n \to R$$

$$X \subseteq R^n.$$

Methods for Solving the Problem

If the objective function f is linear and the constrained space is a polytope, the problem is a linear programming problem, which may be solved using well known linear programming solutions.

If the objective function is concave (maximisation problem), or convex (minimisation problem) and the constraint set is convex, then the program is called convex and general methods from convex optimisation can be used in most cases.

If the objective function is a ratio of a concave and a convex function (in the maximisation case) and the constraints are convex, then the problem can be transformed to a convex optimisation problem using fractional programming techniques.

Several methods are available for solving nonconvex problems. One approach is to use special formulations of linear programming problems. Another method involves the use of branch and bound techniques, where the program is divided into subclasses to be solved with convex (minimisation problem) or linear approximations that form a lower bound on the overall cost within the subdivision. With subsequent divisions, at some point an actual solution will be obtained whose cost is equal to the best lower bound obtained for any of the approximate solutions. This solution is optimal, although possibly not unique. The algorithm may also be stopped early, with the assurance that the best possible solution is within a tolerance from the best point found; such points are called ε-optimal. Terminating to ε-optimal points is typically necessary to ensure finite termination. This is especially useful for large, difficult problems and problems with uncertain costs or values where the uncertainty can be estimated with an appropriate reliability estimation.

Under differentiability and constraint qualifications, the Karush–Kuhn–Tucker (KKT) conditions provide necessary conditions for a solution to be optimal. Under convexity, these conditions are also sufficient.

Examples

2-dimensional Example

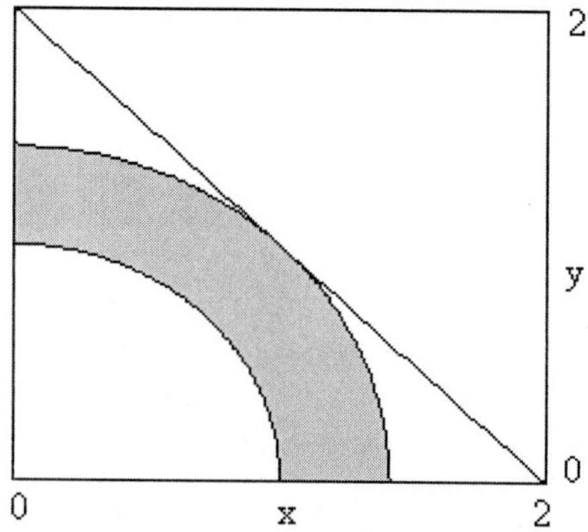

Figure : The intersection of the line with the constrained space represents the solution

A simple problem can be defined by the constraints

$$x_1 \geq 0$$
$$x_2 \geq 0$$
$$x_1^2 + x_2^2 \geq 1$$
$$x_1^2 + x_2^2 \leq 2$$

with an objective function to be maximised

$$f(\text{x}) = x_1 + x_2$$

where x = (x_1, x_2). Solve 2-D Problem.

3-dimensional Example

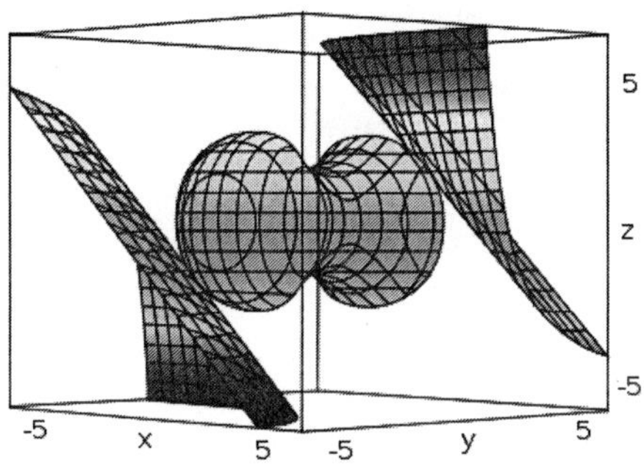

The intersection of the top surface with the constrained space in the center represents the solution

Another simple problem can be defined by the constraints

$$x_1^2 - x_2^2 + x_3^2 \leq 2$$
$$x_1^2 + x_2^2 + x_3^2 \leq 10$$

with an objective function to be maximised

$$f(\text{x}) = x_1x_2 + x_2x_3$$

where x = (x_1, x_2, x_3). Solve 3-D Problem.

Domain decomposition methods

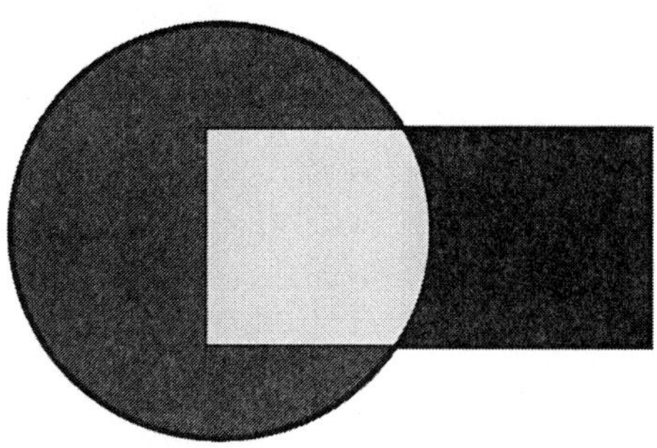

Figure : Domain decomposition methods

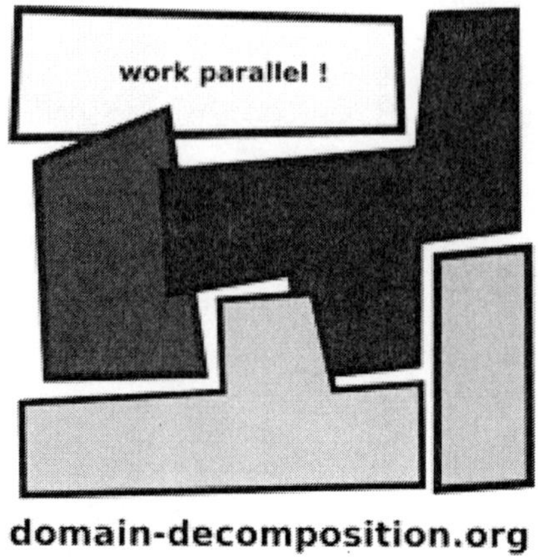

Figure : Domain decomposition methods for non-matching grids

In mathematics, numerical analysis, and numerical partial differential equations, domain decomposition methods solve a boundary value problem by splitting it into smaller boundary value problems on subdomains and iterating to coordinate the solution between adjacent subdomains. A coarse problem with one or few unknowns per subdomain is used to further coordinate the solution between the subdomains globally. The problems on the subdomains are independent, which makes domain decomposition methods suitable for parallel computing. Domain decomposition methods are typically used as preconditioners for Krylov space iterative methods, such as the conjugate gradient method or GMRES.

In overlapping domain decomposition methods, the subdomains overlap by more than the interface. Overlapping domain decomposition methods include the Schwarz alternating method and the additive Schwarz method. Many domain decomposition methods can be written and analysed as a special case of the abstract additive Schwarz method.

In non-overlapping methods, the subdomains intersect only on their interface. In primal methods, such as Balancing domain decomposition and BDDC, the continuity of the solution across subdomain interface is enforced by representing the value of the solution on all neighbouring subdomains by the same unknown. In dual methods, such as FETI, the continuity of the solution across the subdomain interface is enforced by Lagrange multipliers. The FETI-DP method is hybrid between a dual and a primal method.

Non-overlapping domain decomposition methods are also called iterative substructuring methods.

Mortar methods are discretisation methods for partial differential equations, which use separate discretisation on nonoverlapping subdomains. The meshes on the subdomains do not match on the interface, and the equality of the solution is enforced by Lagrange multipliers, judiciously chosen to preserve the accuracy of the solution. In the engineering practice in the finite element method, continuity of solutions between non-matching subdomains is implemented by multiple-point constraints. Finite element simulations of moderate size models require solving linear systems with millions of unknowns. Several hours per time step is an average sequential run time, therefore, parallel computing is a necessity. Domain decomposition methods embody large potential for a parallelisation of the finite element methods, and serve a basis for distributed, parallel computations.

Example 1: 1D Linear BVP

$$u''(x)-u(x)=0$$

$$u(0)=0, u(1)=1$$

The exact solution is:

$$u(x)=\frac{e^{x}-e^{-x}}{e^{1}-e^{-1}}$$

Subdivide the domain into two subdomains, one from $[0,\frac{1}{2}]$ and another from $[\frac{1}{2},1]$. In each of these two subdomains define interpolating functions $v_1(x)$ and $v_2(x)$ At the interface between these two subdomains the following inferface conditions shall be imposed:

$$v_1\left(\frac{1}{2}\right)=v_2\left(\frac{1}{2}\right)$$

$$v_{1'}\left(\frac{1}{2}\right)=v_{2'}\left(\frac{1}{2}\right)$$

Let the interpolating functions be defined as:

$$v_1(x)=\sum_{n=0}^{N}u_n C_n(y_1(x))$$

$$v_2(x)=\sum_{n=0}^{N}u_{n+N} C_n(y_2(x))$$

$$y_1(x) = 4x - 1$$

Where $T_n(y)$ is the nth cardinal function of the chebyshev polynomials of the first kind with input argument y.

If N=4 then the following approximation is obtained by this scheme:

$$u_1 = 0.06236$$

$$u_2 = 0.21495$$

$$u_3 = 0.37428$$

$$u_4 = 0.44341$$

$$u_5 = 0.51492$$

$$u_6 = 0.69972$$

$$u_7 = 0.90645$$

This was obtained with the following MATLAB code.

```
clear all
N=4;
a1=0; b1=1/2;
[T D1 D2 E1 E2 x xsub]=cheb(N,a1,b1); % the diff matrices on
[0,1/2] are the same
%as those on [1/2 1].
I=eye(N+1);
H=D2-I;
H1=[[1 zeros(1,N)]; H(2:end-1,:); [zeros(1,N) 1]];
H1=[H1 [zeros(N,N+1); -[1 zeros(1,N)]]];
H2=[D1(1,:); H(2:end-1,:); [zeros(1,N) 1]];
H2=[[-D1(N+1,:); zeros(N,N+1)] H2];
K=[H1; H2];
F=[zeros(2*N+1,1); 1];
u=K\F;
xx=-cos(pi*(0:N)'/N);
x1=1/4*(xx+1); x2=1/4*(xx+3);
x=[x1; x2];
uex=(exp(x)-exp(-x))./(exp(1)-exp(-1));
```

Decomposition Method (Constraint Satisfaction)

In constraint satisfaction, a decomposition method translates a constraint satisfaction problem into another constraint satisfaction problem that is binary and acyclic. Decomposition methods work by grouping variables into sets, and solving a subproblem for each set. These translations are done because solving binary acyclic problems is a tractable problem.

Each structural restriction defined a measure of complexity of solving the problem after conversion; this measure is called *width*. Fixing a maximal allowed width is a way for identifying a subclass of constraint satisfaction problems. Solving problems in this class is polynomial for most decompositions; if this holds for a decomposition, the class of fixed-width problems form a tractable subclass of constraint satisfaction problems.

Overview

Decomposition methods translate a problem into a new one that is easier to solve. The new problem only contains binary constraints; their scopes form a directed acyclic graph. The variables of the new problem represent each a set of variables of the original one. These sets are not necessarily disjoint, but they cover the set of the original variables. The translation finds all partial solutions relative to each set of variables. The problem that results from the translation represents the interactions between these local solutions.

By definition, a decomposition method produces a binary acyclic problem; such problems can be solved in time polynomial in its size. As a result, the original problem can be solved by first translating it and then solving the resulting problem; however, this algorithm is polynomial-time only if the decomposition does not increase size superpolynomially. The *width* of a decomposition method is a measure of the size of problem it produced. Originally, the width was defined as the maximal cardinality of the sets of original variables; one method, the hypertree decomposition, uses a different measure. Either way, the width of a decomposition is defined so that decompositions of size bounded by a constant do not produce excessively large problems. Instances having a decomposition of fixed width can be translated by decomposition into instances of size bounded by a polynomial in the size of the original instance.

The width of a problem is the width of its minimal-width decomposition. While decompositions of fixed width can be used to efficiently solve a problem, a bound on the width of instances does necessarily produce a tractable structural restriction. Indeed, a fixed width problem has a decomposition of fixed width, but finding it may not be polynomial. In order for a problem of fixed width being efficiently solved by decomposition, one of its decompositions of low width has to be found efficiently. For this reason, decomposition methods and their associated width are defined in such a way not only solving the problem given a fixed-width decomposition of it is polynomial-time, but also finding a fixed width decomposition of a fixed-width problem is polynomial-time.

Decomposition Methods

Decomposition methods create a problem that is easy to solve from an arbitrary one. Each variable of this new problem is associated to a set of original variables; its domain contains tuples of values for the variables in the associated set; in particular, these are the tuples that satisfy a set of constraints over these variables.

The constraints of the new problem bounds the values of two new variables to have as values two tuples that agree on the shared original variables. Three further conditions ensure that the new problem is equivalent to the old one and can be solved efficiently.

In order for the new problem to be solvable efficiently, the primal graph of the new problem is required to be acyclic. In other words, viewing the variables as vertices and the (binary) constraints as edges, the resulting graph is required to be a tree or a forest.

In order for the new problem to be equivalent to the old one, each original constraint is enforced as part of the definition of the domain of at least one new variables. This requires that, for each constraint of the old problem, there exists a variable of the new problem such that its associated set of original variables include the scope of the constraint, and all tuples in its domain satisfy the constraint.

A further condition that is necessary to ensure equivalence is that the binary constraints are sufficient to enforce all "copies" of each original variable to assume the same value. Since the same original variable can be associated to several of the new variables, the values of these new variable must all agree on the value of the old variable.

The binary constraints are used to enforce equality of the old variables shared between the two new variables. Two copies of a new variable are forced equal if there exists a path of binary constraints between their new variables and all new variables in this path contain the old variable.

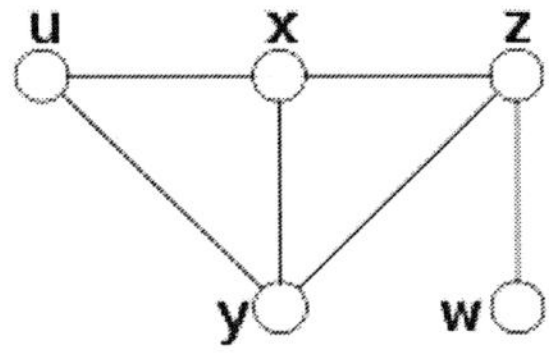

Figure : An example constraint satisfaction problem; this problem is binary, and the constraints are represented by edges of this graph.

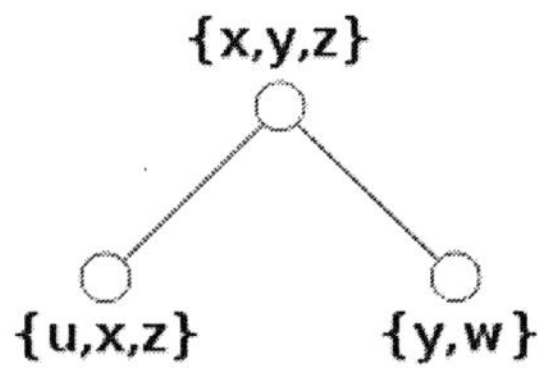

Figure : A decomposition tree; for every edge of the original graph, there is a node that contains both its endpoints; all nodes containing a variable are connected

x	y
1	2
1	3
2	3

x	z
1	3
2	2

y	z
2	3
3	2

⇨

x	y	z
1	2	3
2	3	2

Figure : Solving a subproblem. This example shows solving the subproblem made of all constraints on the variables of the first set *{x,y,z}*. A similar procedure is done for the other sets *{u,x,z}* and {y,w}

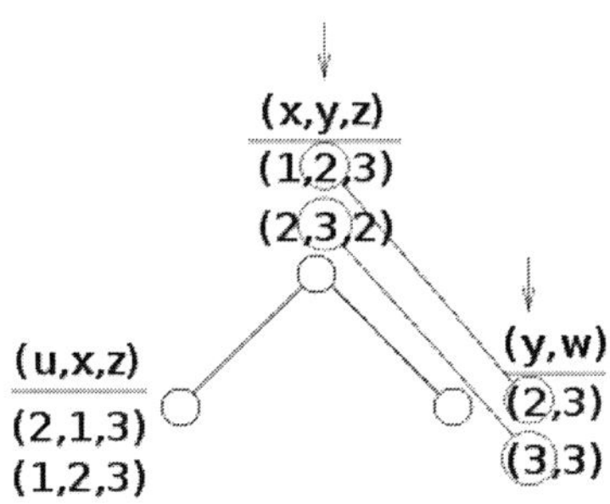

Figure : Each node of the tree is made a variable. Its domain is the set of solutions of the partial problem, which is a set of tuples. The constraints of the new problem allow only the tuples that contain equal values of the original variables. In the figure, equality of *y* is shown: the corresponding constraint is only met by the first tuple of *{x,y,z}* and the first tuple of *(y,w)*, and by the second tuple of *{x,y,z}* and the second tuple of *(y,w)*.

A decomposition method is usually defined by providing a tree whose nodes are the variables of the new problem; for each node, also provided are the associated set of original variables and possibly a set of original constraints used to build the domain of the variable in the new problem.

Of the above three conditions (tree structure, enforcement of constraints, and equivalence of copies of original variables), the first one is automatically enforced by this definition. The condition of enforcement of constraints is in mostly formulated as: the scope of each constraint is a subset of the variables of some node; however, a different condition may be used when nodes are also associated to sets of constraints. The equivalence of all copies of the original variables is usually formulated as: the subgraph induced by the nodes associated to an original variable is connected.

Decomposition Methods for Binary Problems

A number of decomposition methods exist. Most of them generate a tractable class by bounding the width of instances. The following are the decomposition methods defined for binary constraint satisfaction problems. Since a problem can be made binary by translating it into its dual problem or using hidden variables, these methods can be indirectly used to provide a tree decomposition for arbitrary constraint satisfaction problems.

Biconnected Components

In graph theory, a separating vertex is a node of a graph that "breaks" the graph when removed from it. Formally, it is a node whose removal from the graph increases the number of its connected components. A biconnected component of a graph is a maximal set of its nodes whose induced subgraph is connected and does not have any separating vertex.

It is known from graph theory that the biconnected components and the separating vertices of a graph form a tree. This tree can be built as follows: its nodes are the biconnected components and the separating vertices of the graph; edges only connect a biconnected component with a separating vertex, and in particular this happens if the vertex is contained in the component. It can be proved that this graph is actually a tree. If the constraints of a binary constraint satisfaction problem are viewed as edges of a graph whose nodes are the variables, this tree is a decomposition of the problem. The width of a decomposition is the maximal number of vertices in a biconnected component.

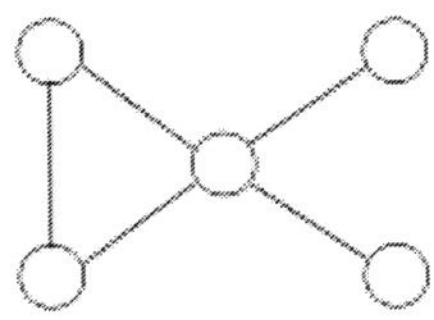

Figure : The primal graph of a constraint satisfaction problem (every node represents a variable and every edge a constraint between two variables)

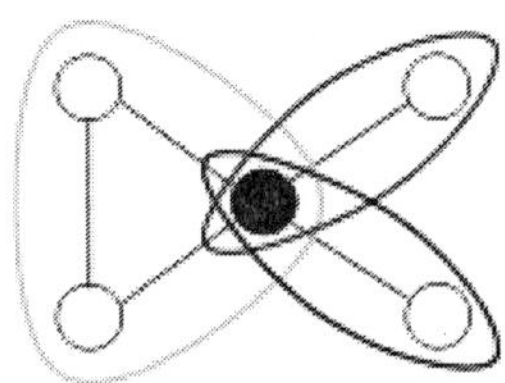

Figure : Its biconnected components (coloured) and its separating vertices (black, only one in this case)

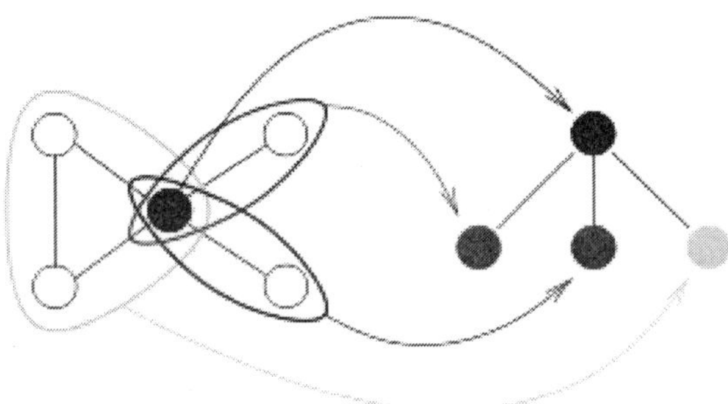

Figure : The biconnected decomposition: nodes of the tree are separating vertices *and* biconnected components

Cycle Cutset

The cycle decomposition method split a problem into a cyclic and an acyclic part. While it does not fit in the definition of the other decomposition methods, which produce a tree whose nodes are labelled with sets of nodes, it can be easily reformulated in such terms. This decomposition method is based on the idea that, after some variables are given a value, what remains of the problem once these variables have been eliminated may be acyclic. Formally, a cycle cutset of a graph is a set of nodes that makes the graph acyclic when they are removed from it. A similar definition can be given for a constraint satisfaction problem using its primal graph.

The width of a cycle decomposition is the number of variables in the cutset. The width of a problem is the minimal width of its cycle cutset decompositions.

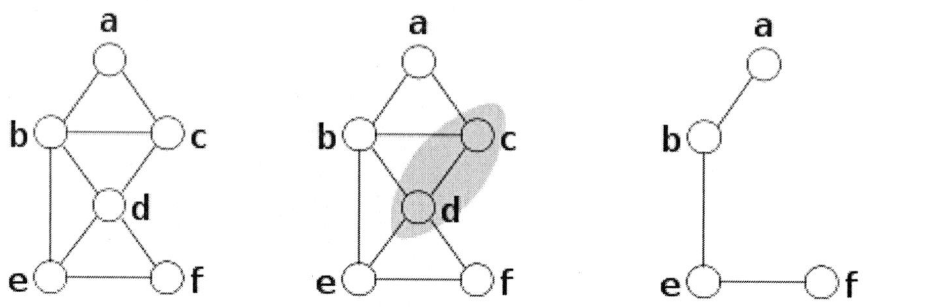

A graph representation of a constraint satisfaction problem

A cycle cutset of the graph (others exist)

Removing the cycle cutset, what remains is an acyclic graph

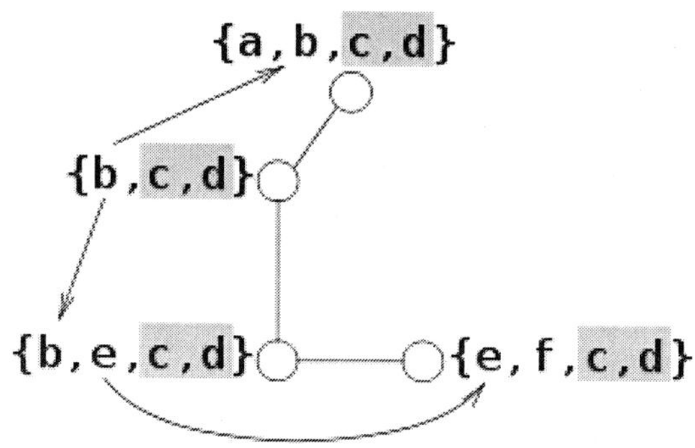

Figure : Choosing the node b as the root, this is the tree similar to the ones created by the other decomposition methods.

Such a decomposition does not fit in the scheme of the other decompositions because the result of the decomposition is not a tree, but rather a set of variables (those of the cutset) and a tree (formed by the variables not in the cutset).

However, a tree like those generated by the other decomposition methods can be obtained from the tree that results from removing the cutset; this is done by choosing a root, adding all variables of the cutset to all its nodes, and the variables of each node to all its children. This results in a tree whose maximal number of variables associated with a node is equal to the size of the cutset plus two. Apart from the addition of two, this is the width of the decomposition, which is defined as the number of variables in the considered cutset.

Tree Decomposition

Tree decomposition is a well-known concept from graph theory. Reformulated in terms of binary constraints, a tree decomposition is a tree whose nodes are associated to sets of variables; the scope of each constraint is contained in set of variables of some node, and the subtree of nodes associated to each variable is connected. This is the most general form of decomposition for binary constraints that follows the scheme outlined above, as the conditions imposed on the tree are only the ones that are necessary to guarantee equivalent of the original and new problem.

The width of such a decomposition is the maximal number of variables associated to the same node minus one. The treewidth of a problem is the minimal width of its tree decompositions.

Bucket elimination can be reformulated as an algorithm working on a particular tree decomposition. In particular, given an ordering of the variables, every variable is associated a bucket containing all constraints such that the variable is the greatest in their scope. Bucket elimination corresponds to the tree decomposition that has a node for each bucket. This node is associated all of its constraints, and corresponds to the set of all variables of these constraints. The parent of a node associated to the bucket of x_i is the node associated to the bucket of x_j, where is the greatest node that is in a constraint with and precedes in the ordering.

Decomposition Methods for Arbitrary Problems

The following methods can be used for translating an arbitrary constraint satisfaction problem, either binary or otherwise. Since they can also be used on binary problems, they can also be used on the result of making constraints binary, either by translating to the dual problem or by using hidden variables.

Some of these methods associate constraints with nodes of the tree, and define width taking into account the number of constraints associated with nodes. This may reduce the width of some problems. For example, a decomposition in which ten variables are associated with each node has width ten; however, if each of these sets of ten variables is the scope of a constraint, each node can be associated that constraint instead, resulting in width one.

Biconnected Components

The biconnected decomposition of an arbitrary constraint satisfaction problem is the biconnected decomposition of its primal graph. Every constraint can be enforced on a node of the tree because each constraint creates a clique on its variables on the primal graph, and a clique is either a biconnected component or a subset of a biconnected component.

Tree Decomposition

A tree decomposition of an arbitrary constraint satisfaction problem is a tree decomposition of its primal graph. Every constraint can be enforced on a node of the tree because each constraint creates a clique on its variables on the primal graph and, for every tree decomposition, the variables of a clique are completely contained in the variables of some node.

Cycle Hypercutset

This is the same mathod of cycle cutset using the definition of cutset for hypergraphs: a cycle hypercutset of a hypergraph is a set of edges (rather than vertices) that makes the hypergraph acyclic when all their vertices are removed. A decomposition can be obtained by grouping all variables of a hypercutset in a single one. This leads to a tree whose nodes are sets of hyperedges. The width of such a decomposition is the maximal size of the sets associated with nodes, which is one if the original problem is acyclic and the size of its minimal hypercutset otherwise. The width of a problem is the minimal width of its decompositions.

Hinge Decomposition

An hinge is a subset of nodes of hypergraph having some properties defined below. An hinge decomposition is based on the sets of variables that are minimal hinges of the hypergraph whose nodes are the variables of the constraint satisfaction problem and whose hyperedges are the scopes of its constraints. The definition of hinge is as follows. Let H be a set of hyperedges. A path w.r.t. H is a sequence of edges such that the intersection of each one with the next one is non-empty and not contained in the nodes of H. A set of edges is connected w.r.t. H if, for each pair of its edges, there is a path w.r.t. H of which the two nodes are the first and the last edge. A connected component w.r.t. H is a maximal set of connected edges.

Hinges are defined for reduced hypergraphs, which are hypergraphs where no hyperedge is contained in another. A set of at least two edges H is a hinge if, for every connected component F w.r.t. H, all nodes in F that are also in H are all contained in a single edge of H.

An hinge decomposition is based on the correspondence between constraint satisfaction problems and hypergraphs. The hypergraph associated with a problem has the variables of the problem as nodes are the scopes of the constraints as hyperedges. An hinge decomposition of a constraint satisfaction problem is a tree whose nodes are some minimal hinges of the hypergraph associated to the problem and such that some other conditions are met. By the correspondence of problems with hypergraphs, a hinge is a set of scopes of constraints, and can therefore be considered as a set of constraints.

The additional conditions of the definition of a hinge decomposition are three, of which the first two ensure equivalence of the original problem with the new one. The two conditions for equivalence are: the scope of each constraint is contained in at least one node of the tree, and the subtree induced by a variable of the original problem is connected. The additional condition is that, if two nodes are joined, then they share exactly one constraint, and the scope of this constraint contains all variables shared by the two nodes.

The maximal number of constraints of a node is the same for all hinge decompositions of the same problem. This number is called the *degree of cyclicity* of the problem or its hingewidth.

Tree Clustering

Tree clustering or join-tree clustering is based on merging constraints in such a way the resulting problem has a join tree, this join tree is the result of the decomposition.

A join tree of a constraint satisfaction problem is a tree in which each node is associated a constraints (and vice versa) and such that the subtree of nodes whose constraint contains a variable is connected. As a result, producing a join tree can be viewed as a particular form of decomposition, where each node of the tree is associated the scope of a constraint.

Not all constraint satisfaction problems have a join tree. However, problems can be modified to acquire a join tree by merging constraints. Tree clustering is based on the fact that a problem has a join tree if and only if its primal graph is chordal and *conformant* with the problem, the latter meaning that the variables of every maximal clique of the primal graph are the scope of a constraint and vice versa. Tree clustering modify an arbitrary problem in such a way these two conditions are met. Chordality is enforced by adding new binary constraints. Conformality is obtained by merging constraints.

In particular, chordality is enforced by adding some "fake" binary constraints to the problem. These are binary constraints satisfied by any pair of values, and are used only to add edges to the primal graph of the problem. In particular, chordality is obtained by adding edges producing the induced graph of the primal graph according to an arbitrary ordering. This procedure is correct because the induced graph is always chordal and is obtained adding edges to the original graph.

Conformality requires that the maximal cliques of the primal graph are exactly the scope of the constraints. While the scope of every original constraint is clique on the primal graph, this clique is not necessarily maximal. Moreover, even if it initially was maximal, enforcing chordality may create a larger clique. Conformality is enforced by merging constraints. In particular, for every maximal clique of the graph resulting from enforcing chordality, a new constraint with the clique as scope is created. The satisfying values of this new constraint are the ones satisfying all original constraints whose scope is contained in the clique. By this transformation, every original constraint is "included" in at least one new constraint. Indeed, the scope of every original constraint is a clique of the primal graph. This clique remains a clique even after enforcing chordality, as this process only introduces new edges. As a result, this clique either is maximal or is contained in a maximal clique.

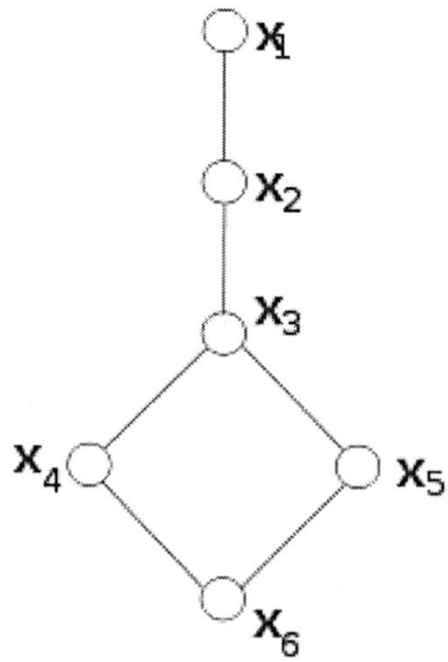

Figure : An example: a binary constraint satisfaction problem (join-tree clustering can also be applied to non-binary constraints.) This graph is not chordal (x3x4x5x6 form a cycle of four nodes having no chord).

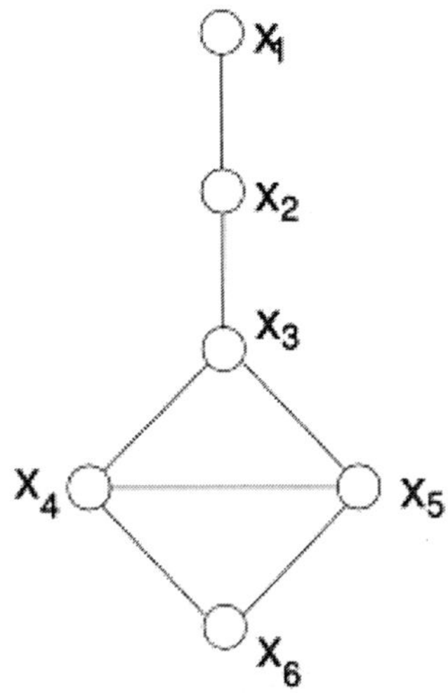

Figure : The graph is made chordal. The algorithm analyses the nodes from x6 to x1. The red edge is the only edge added because x6 is the only node having two non-joined parents. It represents a constraint between x4 and x5 that is satisfied by every pair of values.

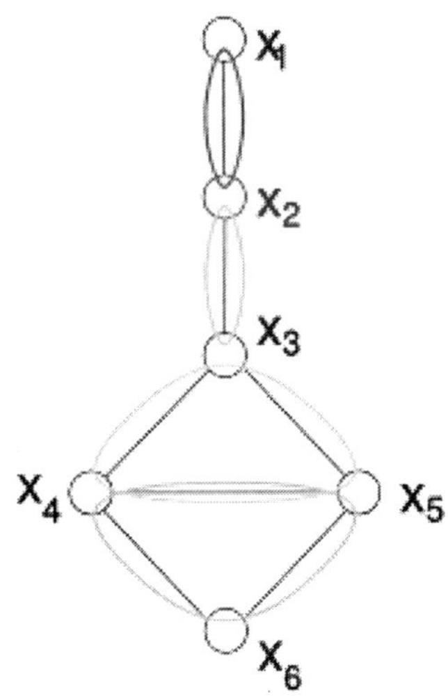

Figure : The maximal cliques of the resulting graph are identified. Join-tree clustering replaces the constraints on {x3, x4, x5, x6} with two equivalent constraints, one on {x3, x4, x5} and one on {x4, x5, x6}.

This translation requires the maximal cliques of a chordal graph to be identified. However, this can bone easily using the same ordering used for enforcing chordality. Since the parents of each node are all connected to each other, the maximal cliques are composed of a node (the maximal node of the clique in a max-cardinality ordering) and all its parents. As a result, these maximal cliques can be detected by considering each node with its parents.

The problem that results from this process has a join tree, and such a join tree can be obtained by using the same ordering of variables again. Proceeding from the last node to the first, every constraint is connected with the preceding constraint that shares more variables with it. Join-tree clustering can be seen as a decomposition method in which:

- the elements of the cover are the cliques of the graph resulting from enforcing chordality;
- the tree is the join tree;
- every constraint is assigned to all nodes of the tree whose sets of variables contain the scope of the constraint.

The width of a tree-clustering decomposition is the maximal number of variables associated with each node of the tree. The width of a problem is the minimal width of its tree-clustering decompositions.

Hinge/Clustering Decomposition

The result of hinge decomposition can be further simplified by decomposing each hinge using tree clustering. In other words, once the hinges have been identified, a tree clustering of each of them is produced. In terms of the resulting tree, each node is therefore replaced by a tree.

Query Decomposition

Query decomposition associates a set of variables and a set of constraints to each node of a tree; each constraint is associated to some node, and the subtree induced by the nodes associated to a given variable or constraint is connected. More precisely, for each variable, the subtree of nodes associated to this variable or with a constraint having this variable in its scope is connected. The width of a decomposition is the maximal combined number of variables and constraints associated with a node.

Associating constraints with nodes possibly reduces the width of decompositions and of instances. On the other hand, this definition of width still allows problems of fixed width to be solved in polynomial time if the decomposition is given. In this case, the domain of a new variable is obtained by solving a subproblem which can be polynomially large but has a fixed number of constraints.

As a result, this domain is guaranteed to be of polynomial size; the constraints of the new problem, being equalities of two domains, are polynomial in size as well.

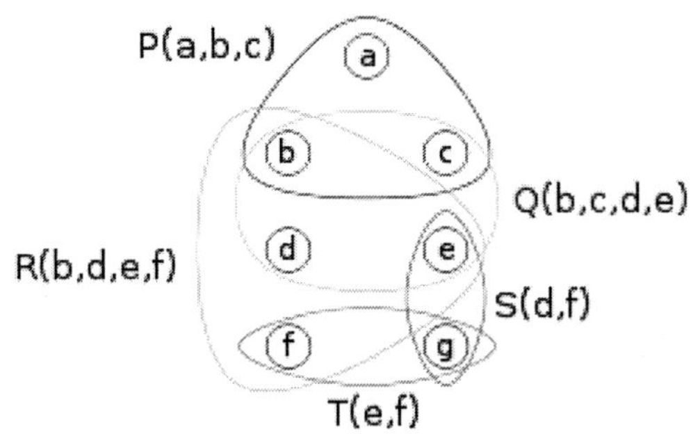

Figure : A hypergraph representation of a constraint satisfaction problem: the constraints are given names (P, Q, R, S, T), and their scopes are shown (P (a, b, c) means that constraint P is on the variables {a, b, c}

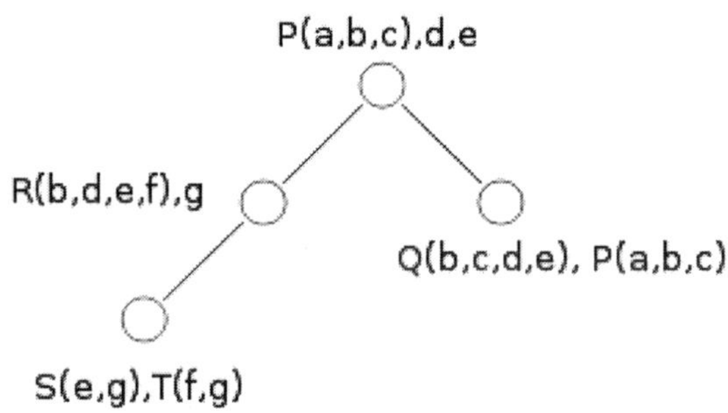

Figure : A query decomposition of the problem. Nodes may contain variables, constraints, or both. Although to the right most node are associated a total of five variables (i.e. a,b,c,d,e among the two constraints), this is a decomposition of width 3 because no node contains more than three constraints and isolated variables (there is another decomposition of width 2 and it is possible to show that this decomposition of width 2 is the minimum width of this hypergraph).

A *pure query decomposition* is a query decomposion in which nodes are only associated to constraints. From a query decomposition of a given width one can build in logarithmic space a pure query decomposition of the same width. This is obtained by replacing the variables of a node that are not in the constraints of the node with some constraints that contain these variables.

A drawback of this decomposition method is that checking whether an instance has a fixed width is in general NP-complete; this has been proved to be the case with width 4.

Hypertree Decomposition

A hypertree decomposition associates a set of variables and a set of constraints to each node of a tree. It extends query decomposition by allowing the constraints of a node to contain variables that are not used when creating the domain of the new variable associated with the node. Beside the common conditions for a decomposition method (the scope of each constraint is in at least a set of variables associated with a node and the subtree induced by an original variable is connected), the following two conditions are required to hold:

1. each original variable in a node is in the scope of at least one constraint associated with the node;
2. the variables of the constraints of a node that are not variables of the node do not occur in the subtree rooted at the node.

The width of a tree decomposition is the maximal number of constraints associated with each node. If this width is bounded by a constant, a problem equivalent to the original one can be built in polynomial time. The variables that are not associated to a node but are in the scope of the constraints of the node are "projected out" when building this instance. This can be done by first projecting the constraints over the variables of the node and then finding all solutions to this subproblem, or by first solving the subproblem with all constraints and then removing the extra variables.

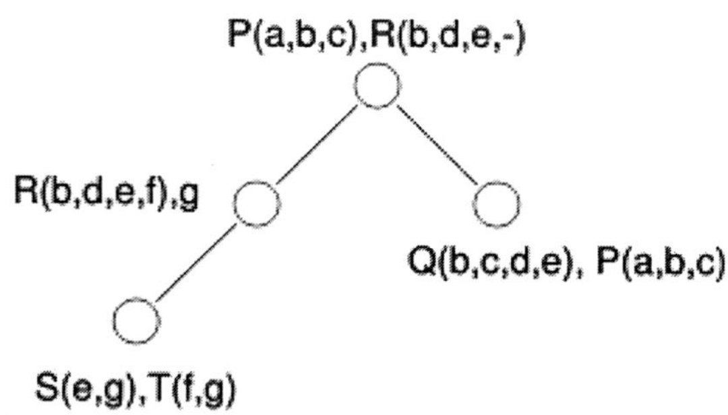

Figure : An hypertree decomposition of the same problem of the query decomposition above. R (b,d,e,-) means that the last variable of R is not a variable associated to the root. By grouping two variables in one constraint in the root, width decreases from three to two

The two requirements above are not necessary to guarantee the equivalence of the original and new problem. They are needed to guarantee that problems of bounded width can be solved in polynomial time.

The possibility of associating a constraint with a node while some of its variables are not effectively associated with the node may produce a width that is less than query width. For example, if a node is associated to $\{C(a,b),c,d,e\}$ in a query decomposition, and a constraint $D(c,d,e,f)$ exists, a hypertree decomposition can associate the same node with constraints $\{C,D\}$ and variables $\{a,b,c,d,e\}$. Since only constraints are computed when checking width, this node has width two. The same node has width four when using query decomposition (one constraint and three variables). This width reduction is possible if two or more variables can be replaced with a single constraint, even if this constraint contains a variable that is not associated with the node.

Generalised Hypertree Decomposition

Generalised hypertree decompositions are defined like hypertree decompositions, but the last requirement is dropped; this is the condition "the variables of the constraints of a node that are not variables of the node do not occur in the subtree rooted at the node". A problem can be clearly solved in polynomial time if a fixed-width decomposition of it is given. However, the restriction to a fixed width is not known to being tractable, as the complexity of finding a decomposition of fixed width even if one is known to exist is not known, as of 2001.

Comparison

The width of instances is a form of efficiency of decomposition methods. Indeed, given that problems can be solved from decompositions of fixed width, the less the width according to a decomposition, the more the instances that can be solved efficiently using that decomposition.

Some decompositions use the number of variables of a node (or a similar amount) as the width. Others do not: cycle hypercutset, hinge decomposition, query decomposition, hypertree decomposition, and generalised hypertree decomposition associate constraints (or their scopes in form of hyperedges) with nodes, and include the number of constraints associated to a node in the width. This can be a significant save in terms of width. Indeed, problems with a single constraint on n variables can only be decomposed in a tree with a single node. This node can be associated with the n variables or with the single constraint. Counting the number of variables leads to width n, while counting the number of constraints leads to width 1.

The comparison between all other decomposition methods is based on generalisation and beating. Generalisation means that every problem having width n according to a method has width bounded by $n + k$ for a fixed k. Beating means that there are classes of problems that have fixed width according to a decomposition method but not according to another. The following are the results for arbitrary problems, where query decomposition is not considered:

- hypertree decomposition generalises and beats all other methods
- hinge decomposition enhanced with tree clustering generalises and beats both hinge decomposition and tree clustering
- tree clustering is equivalent to tree decomposition (on the primal graph)
- both hinge decomposition and tree clustering generalise and beat biconnected components
- cycle cutset (on the primal graph) is generalised and beaten by both cycle hypercutset and tree clustering.

It can also be shown that the tree clustering width is equal to the induced width of the problem plus one. The algorithm of adaptive consistency, which is polynomial for problem of fixed induced width, turns problems into equivalent ones in the same way as the first step of tree clustering.

Solving from a Decomposition

Given the tree of a decomposition, solving can be done by constructing the binary tree-like problem as described above, and solving it. This is a polynomial-time problem, as it can be solved in polynomial time using, for example, an algorithm for enforcing directional arc consistency.

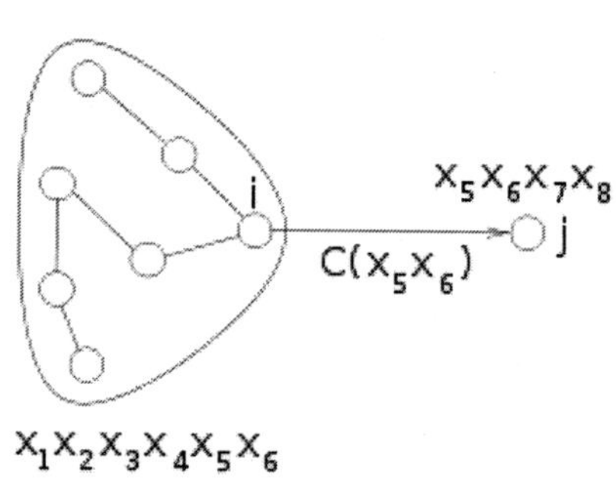

Figure : The constraint passed from node i to node j summarises the effects of the nodes on the "side" of i to the variables of j.

A specialised algorithm for the case of binary acyclic problems that result from a decomposition is described as follows. It works by creating constraints that are passed along the edges of the tree, from the leaves to the root and back. The constraint passed along an edge "summarises" the effects of all constraints of the part of the graph on one side of the edge to the other one.

In a tree, every edge breaks the graph in two parts. The constraint passed along an edge tells how the part of the originating end of the edge affects the variables of the destination node. In other words, a constraint passed from node i to node j tells how the nodes "on the side of i" constrain the variables of node j.

If the variables of these two nodes are X_i and X_j, the nodes on the size of i do not affect all variables X_j but only the shared variables $X_i \cap X_j$. As a result, the influence on of the nodes on the side of can be represented as a constraint on variables $X_i \cap X_j$. Such a constraint can be seen as a "summary" of how a set of nodes affects another node.

The algorithm proceeds from the leaves of the tree. In each node, the summaries of its children (if any) are collected. These summary and the constraint of the node are used to generate the summary of the node for its parent. When the root is reached, the process is inverted: the summary of each node for each child is generated and sent it. When all leaves are reached, the algorithm stops.

Estimation Theory

Estimation theory is a branch of statistics and signal processing that deals with estimating the values of parameters based on measured/ empirical data that has a random component. The parameters describe an underlying physical setting in such a way that their value affects the distribution of the measured data. An estimator attempts to approximate the unknown parameters using the measurements.

For example, it is desired to estimate the proportion of a population of voters who will vote for a particular candidate. That proportion is the unobservable parameter; the estimate is based on a small random sample of voters.Or, for example, in radar the goal is to estimate the range of objects (aeroplanes, boats, etc.) by analysing the two-way transit timing of received echoes of transmitted pulses. Since the reflected pulses are unavoidably embedded in electrical noise, their measured values are randomly distributed, so that the transit time must be estimated.

In estimation theory, it is assumed the measured data is random with probability distribution dependent on the parameters of interest. For example, in electrical communication theory, the measurements which contain information regarding the parameters of interest are often associated with a noisy signal. Without randomness, or noise, the problem would be deterministic and estimation would not be needed.

Estimation Process

The entire purpose of estimation theory is to arrive at an estimator, and preferably an implementable one that could actually be used. The estimator takes the measured data as input and produces an estimate of the parameters.

It is also preferable to derive an estimator that exhibits optimality. Estimator optimality usually refers to achieving minimum average error over some class of estimators, for example, a minimum variance unbiased estimator. In this case, the class is the set of unbiased estimators, and the average error measure is variance (average squared error between the value of the estimate and the parameter). However, optimal estimators do not always exist.

These are the general steps to arrive at an estimator:

- In order to arrive at a desired estimator, it is first necessary to determine a probability distribution for the measured data, and the distribution's dependence on the unknown parameters of interest. Often, the probability distribution may be derived from physical models that explicitly show how the measured data depends on the parameters to be estimated, and how the data is corrupted by random errors or noise. In other cases, the probability distribution for the measured data is simply "assumed", for example, based on familiarity with the measured data and/or for analytical convenience.
- After deciding upon a probabilistic model, it is helpful to find the limitations placed upon an estimator. This limitation, for example, can be found through the Cramér–Rao bound.
- Next, an estimator needs to be developed or applied if an already known estimator is valid for the model. The estimator needs to be tested against the limitations to determine if it is an optimal estimator (if so, then no other estimator will perform better).
- Finally, experiments or simulations can be run using the estimator to test its performance.

After arriving at an estimator, real data might show that the model used to derive the estimator is incorrect, which may require repeating these steps to find a new estimator. A non-implementable or infeasible estimator may need to be scrapped and the process started anew. In summary, the estimator estimates the parameters of a physical model based on measured data.

Basics

To build a model, several statistical "ingredients" need to be known. These are needed to ensure the estimator has some mathematical tractability instead of being based on "good feel".

The first is a set of statistical samples taken from a random vector (RV) of size N. Put into a vector,

$$\mathbf{x} = \begin{bmatrix} x[0] \\ x[1] \\ \vdots \\ x[N-1] \end{bmatrix}.$$

Secondly, we have the corresponding M parameters

$$\boldsymbol{\theta} = \begin{bmatrix} \theta_1 \\ \theta_2 \\ \vdots \\ \theta_M \end{bmatrix},$$

which need to be established with their continuous probability density function (pdf) or its discrete counterpart, the probability mass function (pmf)

$$p(\mathbf{x} \mid \boldsymbol{\theta}).$$

It is also possible for the parameters themselves to have a probability distribution (e.g., Bayesian statistics). It is then necessary to define the Bayesian probability

$$\pi(\boldsymbol{\theta}).$$

After the model is formed, the goal is to estimate the parameters, commonly denoted $\hat{\boldsymbol{\theta}}$, where the "hat" indicates the estimate.

One common estimator is the minimum mean squared error estimator, which utilises the error between the estimated parameters and the actual value of the parameters

$$\mathbf{e} = \hat{\boldsymbol{\theta}} - \boldsymbol{\theta}$$

as the basis for optimality. This error term is then squared and minimised for the MMSE estimator.

Estimators

Commonly-used estimators and estimation methods, and topics related to them:

- Maximum likelihood estimators
- Bayes estimators
- Method of moments estimators
- Cramér–Rao bound
- Minimum mean squared error (MMSE), also known as Bayes least squared error (BLSE)
- Maximum a posteriori (MAP)
- Minimum variance unbiased estimator (MVUE)
- Best linear unbiased estimator (BLUE)
- Unbiased estimators
- Particle filter
- Markov chain Monte Carlo (MCMC)
- Kalman filter
- Ensemble Kalman filter (EnKF)
- Wiener filter.

Chapter 4

Power System Simulation

Power system simulation models are a class of computer simulation programs that focus on the operation of electrical power systems. These computer programs are used in a wide range of planning and operational situations including:

1. Long-term generation and transmission expansion planning
2. Short-term operational simulations
3. Market analysis (e.g. price forecasting)

These programs typically make use of mathematical optimisation techniques such linear programming, quadratic programming, and mixed integer programming.

Key elements of power systems that are modelled include:

1. Load flow (power flow study)
2. Short circuit
3. Transient stability
4. Optimal dispatch of generating units (unit commitment)
5. Transmission (optimal power flow).

Layered on top if this physical framework are models of competition such as Cournot, Bertrand competition, and Supply Function Equilibrium. Before the advent of large scale digital computers, power system simulation was carried out on network analysers, which were essentially miniature scale models of power systems with scaled generators, loads, and line simulators.

Load Flow Calculation

The load-flow calculation is the most common network analysis tool for examining the undisturbed and disturbed network within the scope of operational and strategical planning. On the basis of the network topology with the impedances of all devices as well as with the infeeds and the consumers, the load-flow calculation can provide voltage profiles for all nodes and loading of network components, such as cables and transformers. With this information, compliance to operating limitations such as those stipulated by voltage ranges and maximum loads, can be examined. This is, for example, important for determining the transmission capacity of underground cables, where the influence of cable bundling on the load capability of each cable has to be taken also into account.

Due to the ability to determine losses and reactive-power allocation, load-flow calculation also supports the planning engineer in the investigation of the most economical operation mode of the network.

When changing over from single and/or multi-phase infeed low-voltage meshed networks to isolated networks, load-flow calculation is essential for operational and economical reasons. Load-flow calculation is also the basis of all further network studies, such as motor start-up or investigation of scheduled or unscheduled outages of equipment within the outage simulation. Especially when investigating motor start-up, the load-flow calculation results give helpful hints, for example, of whether the motor can be started in spite of the voltage drop caused by the start-up current.

Short Circuit Analysis

Short circuit analysis analyses the power flow after a fault occurs in a power network. The faults may be three-phase short circuit, one-phase grounded, two-phase short circuit, two-phase grounded, one-phase break, two-phase break or complex faults.

Transient Stability Simulation

The goal of transient stability simulation of power systems is to analyse the stability of a power system in a time window of a few seconds to several tens of seconds. Stability in this aspect is the ability of the system to quickly return to a stable operating condition after being exposed to a disturbace such as for example a tree falling over an overhead line resulting in the automatic disconnection of that line by its protection systems. In engineering terms, a power system is deemed stable if the rotational speeds of motors and generators, and substation voltage levels must return to their normal values in a quick and stable manner.

Unit Commitment

The problem of unit commitment involves finding the least-cost dispatch of available generation resources to meet the electrical load.

Generating resources can include a wide range of types:

1. Nuclear
2. Thermal (using coal, gas, other fossil fuels, or biomass)
3. Renewables (including hydro, wind, wave-power, and solar).

The key decision variables that are decided by the computer program are:

1. Generation level (in megawatts)
2. Number of generating units *on*.

The latter decisions are binary (0,1), which means the mathematical problem is not continuous.

In addition, generating plant are subject to a number of complex technical constraints, including:

1. Minimum stable operating level
2. Maximum rate of ramping up or down
3. Minimum time period the unit is *up* and/or *down*.

These constraints are amenable to mathematical programming as linear or mixed-integer constraints.

Optimal Power Flow

Electricity flows through an AC network according to Kirchhoff's Laws. Transmission lines are subject to thermal limits (simple megawatt limits on flow), as well as voltage and electrical stability constraints. The simulator must calculate the flows in the AC network that result from any given combination of unit commitment and generator megawatt dispatch, and ensure that AC line flows are within both the thermal limits and the voltage and stability constraints. This may include contingencies such as the loss of any one transmission or generation element-a so-called security-constrained optimal power flow (SCOPF), and if the unit commitment is optimised inside this framework we have a security-constrained unit commitment (SCUC).In Optimal Power Flow (OPF) the generalised scalar objective to be minimised is given by:

$$f(u^0,x^0))$$

where u is a set of the control variables, x is a set of independent variables, and the subscript 0 indicates that the variable refers to the pre-contingency power system. The SCOPF is bound by equality and inequality constraint limits. The equality constraint limits are given by the pre and post contingency power flow equations, where k refers to the k'th contingency case:

$g^k(u^k,x^k)=0$ for $k=0,1...n$

The equipment and operating limits are given by the following inequalities:

$U^k_{min} \leq U^k \leq U^k_{max}$ - Represent hard constraints on controls

$X^k_{min} \leq X^k \leq X^k_{max}$ - Represents hard/soft constraints on variables

$h^k(u^k,x^k) \leq 0$ for $k=0,1....n$ - Represents other constraints such as reactive reserve limits

The objective function in OPF can take on different forms relating to active or reactive power quantities that we wish to either minimise or maximise. For example we may wish to minimise transmission losses or minimise real power generation costs on a power network.

Models of Competitive Behaviour

The cost of producing a megawatt of electrical energy is a function of:

1. fuel price
2. generation efficiency (the rate at which potential energy in the fuel is converted to electrical energy)
3. operations and maintenance costs

In addition to this, generating plant incur fixed costs including:

1. plant construction costs, and
2. fixed operations and maintenance costs

Assuming perfect competition, the market-based price of electricity would be based purely on the cost of producing the *next* megawatt of power, the so-called *short-run marginal cost* (SRMC). This price however might not be sufficient to cover the fixed costs of generation, and thus power market prices rarely show purely SRMC pricing. In most established power markets, generators are *free* to offer their generation capacity at prices of their choosing. Competition and use of financial contracts keeps these prices close to SRMC, but inevitably offers price above SRMC do occur (for example during the California energy crisis of 2001). In the context of power system simulation, a number of techniques have been applied to simulate imperfect competition in electrical power markets:

1. Cournot competition
2. Bertrand competition
3. Supply function equilibrium
4. Residual Supply Index analysis.

Various heuristics have also been applied to this problem. The aim is to provide *realistic* forecasts of power market prices, given the forecast supply-demand situation.

Electric Potential Energy

Electric potential energy, or electrostatic potential energy, is a potential energy (measured in joules) that results from conservative Coulomb forces and is associated with the configuration of a particular set of point charges within a defined system. Not to be confused with the term electric potential (measured in volts), the term "electric potential energy" is used to describe the potential energy in systems with electric fields that change with time (time variant), while the term "electrostatic potential energy" is used to describe the potential energy in systems with electric fields that do not change with time (time invariant).

Definition

The reference zero is usually taken to be a state in which the individual point (test) charges are very well separated ("are at infinite separation") and are at rest.

One Point Charge

For one point charge q in the presence of an electric field E due to another point charge Q, the electric potential energy is defined as the negative of the work done to bring it from the reference position r_{ref} to some position r: mathematically this is a line integral. The field is conservative, and for one point charge radial, so it is path independent and equal to the difference in potential between the two endpoints the charge has moved. Mathematically:

$$U_E(r_{\text{ref}}) - U_E(r) = -W_{r_{\text{ref}} \to r} = -\int_{r_{\text{ref}}}^{r} \mathbf{F}\, d\mathbf{r} = -q \int_{r_{\text{ref}}}^{r} \mathbf{E}\, d\mathbf{r}$$

where:

- r = position in 3d space, using cartesian coordinates r = (x, y, z), $r = |\mathrm{r}|$ is the magnitudes of the position vector,
- $W_{r_{\text{ref}} \to r}$ is the work done to bring the charge from the reference position r_{ref} to r,
- F = force exerted on charge q by Q,
- E = electric field due to Q.

Usually U_E is set to zero when r_{ref} is infinity:

$$U_E(r_{\text{ref}} = \infty) = 0$$

so

$$-U_E(r) = -\int_{\infty}^{r} \mathbf{F} \cdot \mathbf{dr} = -q\int_{\infty}^{r} \mathbf{E} \cdot \mathbf{dr}$$

Since E and therefore F, and r, are all radially directed from Q, F and dr must be antiparallel and so

$$\mathbf{F}\,\mathbf{dr} = |\mathbf{F}|\,|\,\mathbf{dr}\,|\cos(\pi) = -F\mathrm{d}r$$

using Coulomb's law:

$$F = \frac{1}{4\pi\varepsilon_0}\frac{qQ}{r^2}$$

allows the integral to easily be evaluated:

$$U_E(r) = \int_{\infty}^{r} \mathbf{F}\,\mathbf{dr} = -\int_{\infty}^{r} \frac{1}{4\pi\varepsilon_0}\frac{qQ}{r^2}\mathrm{d}r = \frac{1}{4\pi\varepsilon_0}\frac{qQ}{r}$$

Sometimes the factor k_e called Coulomb's constant is used in these expressions. In SI units, the Coulomb constant is given by

$$k_e = \frac{1}{4\pi\varepsilon_0},$$

in turn ε_0 is the electric constant.

Many Point Charges

The electric potential energy of a system, U_E, relative to the chosen zero of potential (position where potential is zero), is defined equal to the total work *W* that must be done by a hypothetical external agent to bring the charges slowly, one by one, from infinite separation, to the stated system configuration.

In this process the external agent provides or absorbs any relevant work, and the point charge being slowly moved gains no kinetic energy. Sometimes reference is made to the potential energy of a charge in an electrostatic field. This actually refers to the potential energy of the system containing the charge and the *other* charges that created the electrostatic field.

Electrical energy is energy newly derived from electrical potential energy. When loosely used to describe energy absorbed or delivered by an electrical circuit (for example, one provided by an electric power utility) "electrical energy" refers to energy which has been converted *from* electrical potential energy. This energy is supplied by the combination of electric current and electrical potential that is delivered by the circuit. At the point that this electrical potential energy has been converted to another type of energy, it ceases to be electrical potential energy. Thus, all electrical energy is potential energy before it is delivered to the end-use. Once converted from potential energy, electrical energy can always be described as another type of energy (heat, light, motion, etc.).

To calculate the work required to bring a point charge into the vicinity of other (stationary) point charges, it is sufficient to know only (a) the total field generated by the other charges and (b) the charge of the point charge being moved. The field due to the charge being moved and the values of the other charges are not required. Nonetheless, in many circumstances it is mathematically easier to add up all the pairwise potential energies (as below).

Electrostatic Potential Energy Stored in a Configuration of Discrete Point Charges

The mutual electrostatic potential energy U_E of two charges is equal to the potential energy of a charge in the electrostatic potential generated by the other. That is to say, if charge q_1 generates an electrostatic potential Φ_1, which is a function of position r, then

$$U_E = q_2 \Phi_1(\mathbf{r}_2).$$

Doing the same calculation with respect to the other charge, we obtain

$$U_E = q_1 \Phi_2(\mathbf{r}_1).$$

This can be generalised to give an expression for a group of N charges q_1, q_2 ...q_N at positions r_1, r_2...r_N respectively:

$$U_E = \frac{1}{2}\sum_{i=1}^{N} q_i \Phi(\mathbf{r}_i)$$

where, for each i value, $\Phi(r_i)$ is the electrostatic potential due to all point charges except the one at r_i.

NB: The factor of one half accounts for the 'double counting' of charge pairs. For example, consider the case of just two charges.

One Point Charge

The electrostatic potential energy of a system containing only one point charge is zero, as there are no other sources of electrostatic potential against which an external agent must do work in moving the point charge from infinity to its final location. One should carefully consider the possibility of the point charge interacting with its own electrostatic potential.

However, since such a potential at the location of the point charge itself is infinite, this "self-energy" is intentionally excluded from an evaluation of the total (finite) electrostatic potential energy of the system. Moreover, one may argue that since the electrostatic potential due to the point charge itself provides no work in moving the point charge around, this interaction is unimportant for most purposes.

Two Point Charges

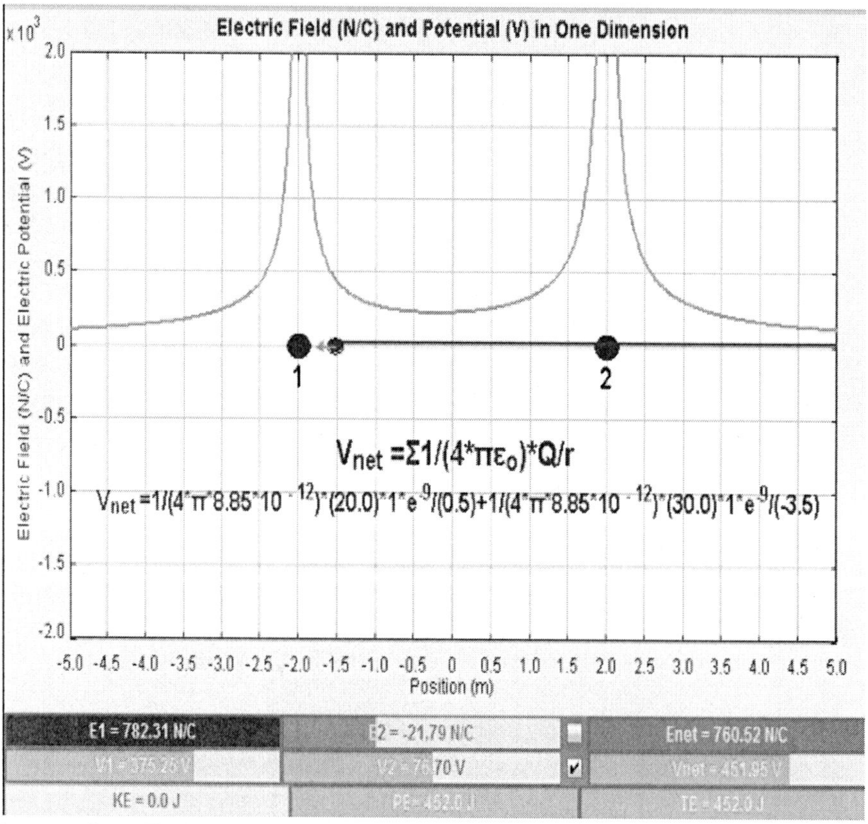

Figure : In a Two point charges system, Electric potential energy U_E of q in the potential well created by Q_1 is $U_E = \dfrac{1}{4\pi\varepsilon_0}\dfrac{qQ_1}{r}$

Consider bringing a second point charge, q, into its final position in the vicinity of the first point charge, Q_1. The electrostatic potential $\Phi(r)$ due to Q_1 is

$$\Phi(r) = k_e \frac{Q_1}{r}$$

Hence we obtain, potential energy of q in the potential of Q as

$$U_E = \frac{1}{4\pi\varepsilon_0} \frac{qQ_1}{r_1}$$

where r_1 is the separation between the two point charges.

Three or more Point Charges

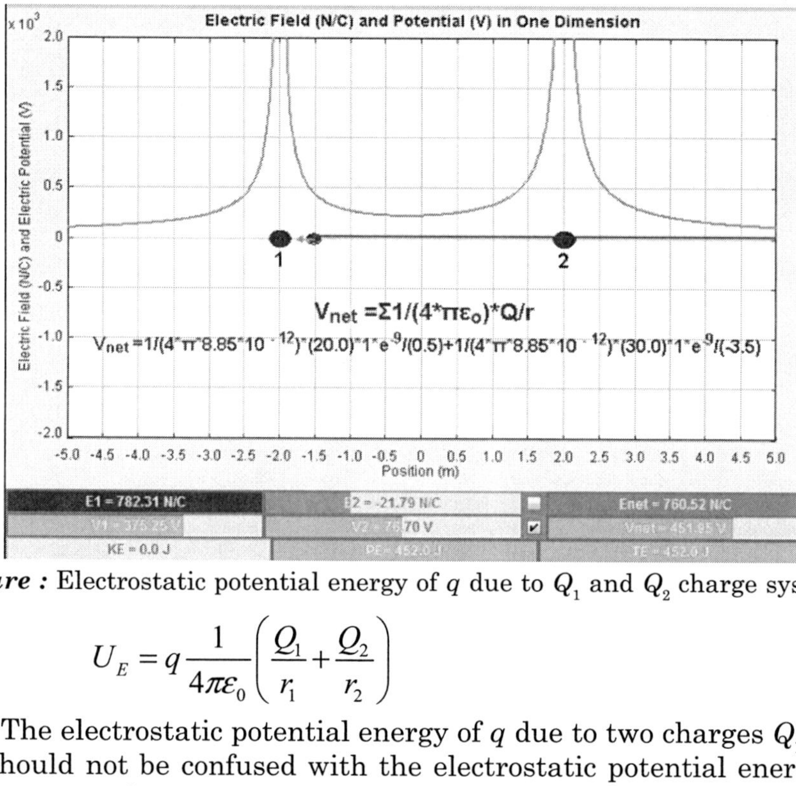

Figure : Electrostatic potential energy of q due to Q_1 and Q_2 charge system :

$$U_E = q\frac{1}{4\pi\varepsilon_0}\left(\frac{Q_1}{r_1} + \frac{Q_2}{r_2}\right)$$

The electrostatic potential energy of q due to two charges Q_1 and Q_2 should not be confused with the electrostatic potential energy of the system of three charges, because the latter also includes the electrostatic potential energy of the system of the two charges Q_1 and Q_2. For three or more point charges, the electrostatic potential energy of the system may be calculated by the total amount of work done by an external agent in bringing individual point charges into their final positions one after another.

NB: Here, ε_0 is the relative permittivity of free space. When the charge is in a medium other than free space or air, the relative permittivity

$$\varepsilon = k\varepsilon_0$$

has to be accounted for where k is the dielectric constant of the medium.

Energy Stored in an Electrostatic Field Distribution

One may take the equation for the electrostatic potential energy of a continuous charge distribution and put it in terms of the electrostatic field. Since Gauss' law for electrostatic field in differential form states

$$\nabla \mathbf{E} = \frac{\rho}{\varepsilon_0}$$

where

- E is the electric field vector
- ρ is the total charge density including dipole charges bound in a material,

then,

$$U = \frac{1}{2} \int_{\text{all space}} \rho(r)\Phi(r) d^3 r$$

$$= \frac{1}{2} \int_{\text{all space}} \varepsilon_0 (\nabla \mathbf{E})\Phi(r) d^3 r$$

so, now using the following divergence vector identity

$$\nabla (\mathbf{A}B) = (\nabla \cdot \mathbf{A})B + \mathbf{A}(\nabla B) \Rightarrow (\nabla \cdot \mathbf{A})B = \nabla (\mathbf{A}B) - \mathbf{A}(\nabla B)$$

we have

$$U = \frac{\varepsilon_0}{2} \int_{\text{all space}} \nabla (\mathbf{E}\Phi) d^3 r - \frac{\varepsilon_0}{2} \int_{\text{all space}} (\nabla \Phi)\, \mathbf{E} d^3 r$$

using the divergence theorem and taking the area to be at infinity where $\Phi(\infty) = 0$

$$U = \frac{\varepsilon_0}{2} \int \Phi \mathbf{E}\, dA - \frac{\varepsilon_0}{2} \int_{\text{all space}} (-\mathbf{E})\, \mathbf{E} d^3 r$$

$$= \int_{\text{all space}} \frac{1}{2} \varepsilon_0 |\mathbf{E}|^2 d^3 r.$$

So, the energy density, or energy per unit volume of the electrostatic field is:

$$u_e = \frac{1}{2}\varepsilon_0 |\mathbf{E}|^2 .$$

Energy in Electronic Elements

Some elements in a circuit can convert energy from one form to another. For example, a resistor converts electrical energy to heat, and a capacitor stores it in its electric field.

The total electric potential energy stored in a capacitor is given by

$$U_E = \frac{1}{2}QV = \frac{1}{2}CV^2 = \frac{Q^2}{2C}$$

where C is the capacitance, V is the electric potential difference, and Q the charge stored in the capacitor.

Electrical Network Frequency Analysis

Electrical network frequency (ENF) analysis is a forensic science technique for validating audio recordings by comparing frequency changes in background mains hum in the recording with long-term high-precision historical records of mains frequency changes from a database. In effect the mains hum signal is treated as if it were a time-dependent digital watermark that can help identify when the recording was created, and help detect any edits in the recording. Historical records of main frequency changes are kept on record f.e. by police in the german federal state of Bavaria since 2010.

The technology has been hailed as "the most significant development in audio forensics since Watergate." However, according to a paper by Huijbregtse and Geradts, the ENF technique, although powerful, has significant limitations caused by ambiguity based on fixed frequency offsets during recording, and self-similarity within the mains frequency database, particularly for recordings shorter than 10 minutes.

More recently, researchers demonstrated that the indoor lights such as fluorescent lights and incandescent bulbs vary their light intensity in accordance with the voltage supplied, which in turn depends on the voltage supply frequency. As a result, the light intensity can carry the frequency fluctuation information to the visual sensor recordings in the similar way as the electromagnetic waves from the power transmission lines carry the ENF information to audio sensing mechanisms. Based on this result, researchers demonstrated that visual track from still video taken in an indoor lighting environments also contain ENF traces that can be extracted by estimating the frequency at which ENF will appear in a video as low sampling frequency of video (25fps-30fps) cause significant aliasing. It was also demonstrated in the same research that the ENF signatures from visual stream and the ENF signature from audio stream in a given video should match. As a result, the matching between the two signal can be used to determine if the audio and visual track was recorded together or superimposed later.

Genetic Algorithm

In the computer science field of artificial intelligence, a genetic algorithm (GA) is a search heuristic that mimics the process of natural evolution. This heuristic is routinely used to generate useful solutions to optimisation and search problems. Genetic algorithms belong to the larger class of evolutionary algorithms (EA), which generate solutions to optimisation problems using techniques inspired by natural evolution, such as inheritance, mutation, selection, and crossover.

Methodology

In a genetic algorithm, a population of strings (called chromosomes or the genotype of the genome), which encode candidate solutions (called individuals, creatures, or phenotypes) to an optimisation problem, evolves toward better solutions. Traditionally, solutions are represented in binary as strings of 0s and 1s, but other encodings are also possible. The evolution usually starts from a population of randomly generated individuals and happens in generations. In each generation, the fitness of every individual in the population is evaluated, multiple individuals are stochastically selected from the current population (based on their fitness), and modified (recombined and possibly randomly mutated) to form a new population. The new population is then used in the next iteration of the algorithm. Commonly, the algorithm terminates when either a maximum number of generations has been produced, or a satisfactory fitness level has been reached for the population. If the algorithm has terminated due to a maximum number of generations, a satisfactory solution may or may not have been reached.

Genetic algorithms find application in bioinformatics, phylogenetics, computational science, engineering, economics, chemistry, manufacturing, mathematics, physics and other fields.

A typical genetic algorithm requires:

1. a genetic representation of the solution domain,
2. a fitness function to evaluate the solution domain.

A standard representation of the solution is as an array of bits. Arrays of other types and structures can be used in essentially the same way. The main property that makes these genetic representations convenient is that their parts are easily aligned due to their fixed size, which facilitates simple crossover operations. Variable length representations may also be used, but crossover implementation is more complex in this case. Tree-like representations are explored in genetic programming and graph-form representations are explored in evolutionary programming.

The fitness function is defined over the genetic representation and measures the *quality* of the represented solution. The fitness function is always problem dependent. For instance, in the knapsack problem one wants to maximise the total value of objects that can be put in a knapsack of some fixed capacity. A representation of a solution might be an array of bits, where each bit represents a different object, and the value of the bit (0 or 1) represents whether or not the object is in the knapsack. Not every such representation is valid, as the size of objects may exceed the capacity of the knapsack. The *fitness* of the solution is the sum of values of all objects in the knapsack if the representation is valid, or 0 otherwise. In some problems, it is hard or even impossible to define the fitness expression; in these cases, interactive genetic algorithms are used.

Once the genetic representation and the fitness function are defined, a GA proceeds to initialise a population of solutions (usually randomly) and then to improve it through repetitive application of the mutation, crossover, inversion and selection operators.

Initialisation

Initially many individual solutions are (usually) randomly generated to form an initial population. The population size depends on the nature of the problem, but typically contains several hundreds or thousands of possible solutions. Traditionally, the population is generated randomly, allowing the entire range of possible solutions (the *search space*). Occasionally, the solutions may be "seeded" in areas where optimal solutions are likely to be found.

Selection (Genetic Algorithm)

Selection is the stage of a genetic algorithm in which individual genomes are chosen from a population for later breeding (recombination or crossover).

A generic selection procedure may be implemented as follows:

1. The fitness function is evaluated for each individual, providing fitness values, which are then normalised. Normalisation means dividing the fitness value of each individual by the sum of all fitness values, so that the sum of all resulting fitness values equals 1.
2. The population is sorted by descending fitness values.
3. Accumulated normalised fitness values are computed (the accumulated fitness value of an individual is the sum of its own fitness value plus the fitness values of all the previous individuals). The accumulated fitness of the last individual should be 1 (otherwise something went wrong in the normalisation step).
4. A random number R between 0 and 1 is chosen.
5. The selected individual is the first one whose accumulated normalised value is greater than R.

If this procedure is repeated until there are enough selected individuals, this selection method is called fitness proportionate selection or *roulette-wheel selection*. If instead of a single pointer spun multiple times, there are multiple, equally spaced pointers on a wheel that is spun once, it is called stochastic universal sampling. Repeatedly selecting the best individual of a randomly chosen subset is tournament selection. Taking the best half, third or another proportion of the individuals is truncation selection.

There are other selection algorithms that do not consider all individuals for selection, but only those with a fitness value that is higher than a given (arbitrary) constant. Other algorithms select from a restricted pool where only a certain percentage of the individuals are allowed, based on fitness value.

Retaining the best individuals in a generation unchanged in the next generation, is called *elitism* or *elitist selection*. It is a successful (slight) variant of the general process of constructing a new population.

Crossover (Genetic Algorithm)

In genetic algorithms, crossover is a genetic operator used to vary the programming of a chromosome or chromosomes from one generation to the next. It is analogous to reproduction and biological crossover, upon which genetic algorithms are based. Cross over is a process of taking more than one parent solutions and producing a child solution from them. There are methods for selection of the chromosomes. Those are also given below.

Methods of Selection of Chromosomes for Crossover

- Roulette wheel selection

It is also known as fitness proportionate selection. The individual is selected on the basis of fitness. The probability of an individual to be selected increases with the fitness of the individual greater or less than its competitor's fitness.

- Boltzmann selection
- Tournament selection
- Rank selection
- Steady state selection.

Crossover Techniques

Many crossover techniques exist for organisms which use different data structures to store themselves.

One-point Crossover

A single crossover point on both parents' organism strings is selected. All data beyond that point in either organism string is swapped between the two parent organisms. The resulting organisms are the children:

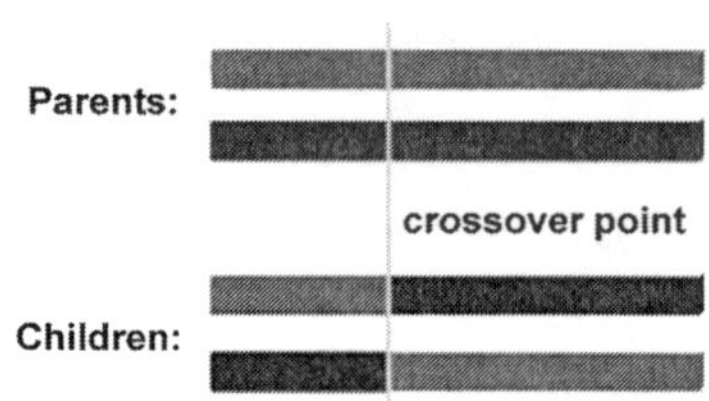

Two-point Crossover

Two-point crossover calls for two points to be selected on the parent organism strings. Everything between the two points is swapped between the parent organisms, rendering two child organisms:

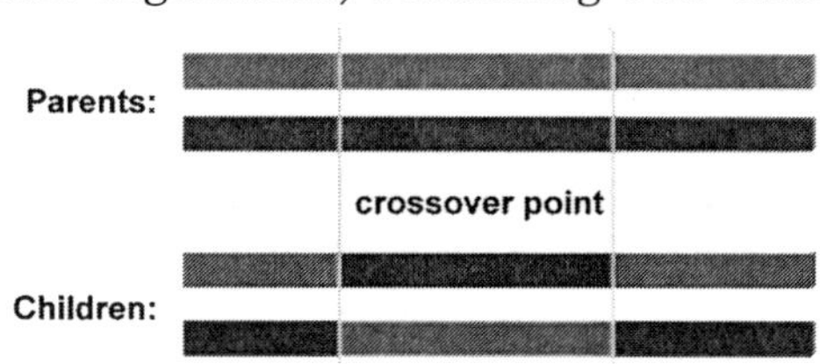

"Cut and Splice"

Another crossover variant, the "cut and splice" approach, results in a change in length of the children strings. The reason for this difference is that each parent string has a separate choice of crossover point.

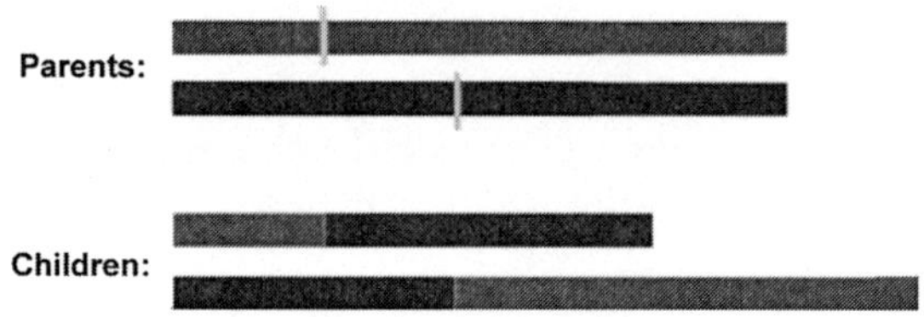

Uniform Crossover and Half Uniform Crossover

The Uniform Crossover uses a fixed mixing ratio between two parents. Unlike one- and two-point crossover, the Uniform Crossover enables the parent chromosomes to contribute the gene level rather than the segment level. If the mixing ratio is 0.5, the offspring has approximately half of the genes from first parent and the other half from second parent, although cross over points can be randomly chosen as seen below

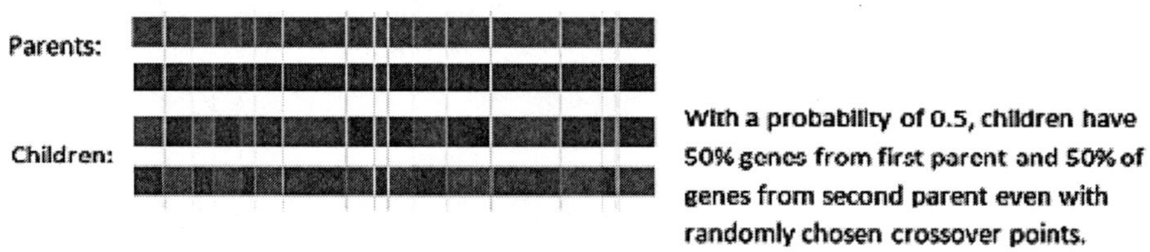

The Uniform Crossover evaluates each bit in the parent strings for exchange with a probability of 0.5. Even though the uniform crossover is a poor method, empirical evidence suggest that it is a more exploratory approach to crossover than the traditional exploitative approach that maintains longer schemata. This results in a more complete search of the design space with maintaining the exchange of good information. Unfortunately, no satisfactory theory exists to explain the discrepancies between the Uniform Crossover and the traditional approaches.

In the uniform crossover scheme (UX) individual bits in the string are compared between two parents. The bits are swapped with a fixed probability, typically 0.5.

In the half uniform crossover scheme (HUX), exactly half of the nonmatching bits are swapped. Thus first the Hamming distance (the number of differing bits) is calculated. This number is divided by two. The resulting number is how many of the bits that do not match between the two parents will be swapped.

Three Parent Crossover

In this technique, the child is derived from three parents. They are randomly chosen. Each bit of first parent is checked with bit of second parent whether they are same. If same then the bit is taken for the offspring otherwise the bit from the third parent is taken for the offspring.

parent1 1 1 0 1 0 0 0 1 0

parent2 0 1 1 0 0 1 0 0 1

parent3 1 1 0 1 1 0 1 0 1

off spring 1 1 0 1 0 0 0 0 1

Crossover for Ordered Chromosomes

Depending on how the chromosome represents the solution, a direct swap may not be possible.

One such case is when the chromosome is an ordered list, such as an ordered list of the cities to be travelled for the travelling salesman problem. There are many crossover methods for ordered chromosomes. The already mentioned N-point crossover can be applied for ordered chromosomes also, but this always need a corresponding repair process, actually, some ordered crossover methods are derived from the idea. However, sometimes a crossover of chromosomes produces recombinations which violate the constraint of ordering and thus need to be repaired. Several examples for crossover operators (also mutation operator) preserving a given order are given in :

1. partially-matched crossover (PMX): In this method, two crossover points are selected at random and PMX proceeds by position wise exchanges. The two crossover points give matching selection. It affects cross by position-by-position exchange operations. In this method parents are mapped to each other hence we can also call it partially mapped crossover.
2. cycle crossover (CX): Beginning at any gene i in parent 1, the -th gene in parent 2 becomes replaced by it. The same is repeated for the displaced gene until the gene which is equal to the first inserted gene becomes replaced (cycle).
3. order crossover operator (OX1): A portion of one parent is mapped to a portion of the other parent. From the replaced portion on, the rest is filled up by the remaining genes, where already present genes are omitted and the order is preserved.
4. order-based crossover operator (OX2)

5. position-based crossover operator (POS)
6. voting recombination crossover operator (VR)
7. alternating-position crossover operator (AP)
8. sequential constrictive crossover operator (SCX)

Other possible methods include the edge recombination operator and partially mapped crossover.

Crossover Biases

For crossover operators which exchange contiguous sections of the chromosomes (e.g. k-point) the ordering of the variables may become important. This is particularly true when good solutions contain building blocks which might be disrupted by a non-respectful crossover operator.

Mutation (Genetic Algorithm)

In genetic algorithms of computing, mutation is a genetic operator used to maintain genetic diversity from one generation of a population of algorithm chromosomes to the next. It is analogous to biological mutation. Mutation alters one or more gene values in a chromosome from its initial state. In mutation, the solution may change entirely from the previous solution. Hence GA can come to better solution by using mutation. Mutation occurs during evolution according to a user-definable mutation probability. This probability should be set low. If it is set to high, the search will turn into a primitive random search.

The classic example of a mutation operator involves a probability that an arbitrary bit in a genetic sequence will be changed from its original state. A common method of implementing the mutation operator involves generating a random variable for each bit in a sequence.

This random variable tells whether or not a particular bit will be modified. This mutation procedure, based on the biological point mutation, is called single point mutation. Other types are inversion and floating point mutation. When the gene encoding is restrictive as in permutation problems, mutations are swaps, inversions and scrambles.

The purpose of mutation in GAs is preserving and introducing diversity. Mutation should allow the algorithm to avoid local minima by preventing the population of chromosomes from becoming too similar to each other, thus slowing or even stopping evolution. This reasoning also explains the fact that most GA systems avoid only taking the fittest of the population in generating the next but rather a random (or semi-random) selection with a weighting toward those that are fitter.

For different genome types, different mutation types are suitable:

Bit string mutation : The mutation of bit strings ensue through bit flips at random positions.

Example:

1	0	1	0	0	1	0
				↓		
1	0	1	0	1	1	0

The probability of a mutation of a bit is $\frac{1}{l}$, where l is the length of the binary vector. Thus, a mutation rate of 1 per mutation and individual selected for mutation is reached.

Flip Bit : This mutation operator takes the chosen genome and inverts the bits. (i.e. if the genome bit is 1, it is changed to 0 and vice versa)

Boundary : This mutation operator replaces the genome with either lower or upper bound randomly. This can be used for integer and float genes.

Non-Uniform : The probability that amount of mutation will go to 0 with the next generation is increased by using non-uniform mutation operator. It keeps the population from stagnating in the early stages of the evolution. It tunes solution in later stages of evolution. This mutation operator can only be used for integer and float genes.

Uniform : This operator replaces the value of the chosen gene with a uniform random value selected between the user-specified upper and lower bounds for that gene. This mutation operator can only be used for integer and float genes.

Gaussian : This operator adds a unit Gaussian distributed random value to the chosen gene. If it falls outside of the user-specified lower or upper bounds for that gene, the new gene value is clipped. This mutation operator can only be used for integer and float genes.

Termination

This generational process is repeated until a termination condition has been reached. Common terminating conditions are:

- A solution is found that satisfies minimum criteria
- Fixed number of generations reached
- Allocated budget (computation time/money) reached
- The highest ranking solution's fitness is reaching or has reached a plateau such that successive iterations no longer produce better results
- Manual inspection
- Combinations of the above.

Simple generational genetic algorithm procedure:

- Choose the initial population of individuals
- Evaluate the fitness of each individual in that population
- Repeat on this generation until termination (time limit, sufficient fitness achieved, etc.):
  - Select the best-fit individuals for reproduction
  - Breed new individuals through crossover and mutation operations to give birth to offspring
  - Evaluate the individual fitness of new individuals
  - Replace least-fit population with new individuals.

The Building Block Hypothesis

Genetic algorithms are simple to implement, but their behaviour is difficult to understand. In particular it is difficult to understand why these algorithms frequently succeed at generating solutions of high fitness when applied to practical problems. The building block hypothesis (BBH) consists of:

1. A description of a heuristic that performs adaptation by identifying and recombining "building blocks", i.e. low order, low defining-length schemata with above average fitness.
2. A hypothesis that a genetic algorithm performs adaptation by implicitly and efficiently implementing this heuristic.

Goldberg describes the heuristic as follows:

> *"Short, low order, and highly fit schemata are sampled, recombined [crossed over], and resampled to form strings of potentially higher fitness. In a way, by working with these particular schemata [the building blocks], we have reduced the complexity of our problem; instead of building high-performance strings by trying every*

conceivable combination, we construct better and better strings from the best partial solutions of past samplings.

"Because highly fit schemata of low defining length and low order play such an important role in the action of genetic algorithms, we have already given them a special name: building blocks. Just as a child creates magnificent fortresses through the arrangement of simple blocks of wood, so does a genetic algorithm seek near optimal performance through the juxtaposition of short, low-order, high-performance schemata, or building blocks."

Observations

There are several general observations about the generation of solutions specifically via a genetic algorithm:

- Selection is clearly an important genetic operator, but opinion is divided over the importance of crossover versus mutation. Some argue that crossover is the most important, while mutation is only necessary to ensure that potential solutions are not lost. Others argue that crossover in a largely uniform population only serves to propagate innovations originally found by mutation, and in a non-uniform population crossover is nearly always equivalent to a very large mutation (which is likely to be catastrophic). There are many references in Fogel (2006) that support the importance of mutation-based search.
- As with all current machine learning problems it is worth tuning the parameters such as mutation probability, crossover probability and population size to find reasonable settings for the problem class being worked on. A very small mutation rate may lead to genetic drift (which is non-ergodic in nature). A recombination rate that is too high may lead to premature convergence of the genetic algorithm. A mutation rate that is too high may lead to loss of good solutions unless there is elitist selection. There are theoretical but not yet practical upper and lower bounds for these parameters that can help guide selection.
- Often, GAs can rapidly locate *good* solutions, even for large search spaces.

Criticisms

There are several criticisms of the use of a genetic algorithm compared to alternative optimisation algorithms:

- Repeated fitness function evaluation for complex problems is often the most prohibitive and limiting segment of artificial evolutionary algorithms. Finding the optimal solution to complex high dimensional, multimodal problems often requires very expensive fitness function evaluations. In real world problems such as structural optimisation problems, one single function evaluation may require several hours to several days of complete simulation. Typical optimisation methods can not deal with such types of problem. In this case, it may be necessary to forgo an exact evaluation and use an approximated fitness that is computationally efficient. It is apparent that amalgamation of approximate models may be one of the most promising approaches to convincingly use GA to solve complex real life problems.
- Genetic algorithms do not scale well with complexity. That is, where the number of elements which are exposed to mutation is large there is often an exponential increase in search space size. This makes it extremely difficult to use the technique on problems such as designing an engine, a house or plane. In order to make such problems tractable to evolutionary search, they must be broken down into the simplest representation possible. Hence we typically see evolutionary algorithms encoding designs for fan blades instead of engines, building shapes instead of detailed construction plans, aerofoils instead of whole aircraft designs. The second problem of complexity is the issue of how to protect parts that have evolved to represent good solutions from further destructive mutation, particularly when their fitness assessment requires them to combine well with other parts. It has been suggested by some in the community that a developmental approach to evolved solutions could overcome some of the issues of protection, but this remains an open research question.
- The "better" solution is only in comparison to other solutions. As a result, the stop criterion is not clear in every problem.
- In many problems, GAs may have a tendency to converge towards local optima or even arbitrary points rather than the global optimum of the problem. This means that it does not "know how" to sacrifice short-term fitness to gain longer-term fitness. The likelihood of this occurring depends on the shape of the fitness landscape: certain problems may provide an easy ascent towards a global optimum, others may make it easier

for the function to find the local optima. This problem may be alleviated by using a different fitness function, increasing the rate of mutation, or by using selection techniques that maintain a diverse population of solutions, although the No Free Lunch theorem proves that there is no general solution to this problem. A common technique to maintain diversity is to impose a "niche penalty", wherein, any group of individuals of sufficient similarity (niche radius) have a penalty added, which will reduce the representation of that group in subsequent generations, permitting other (less similar) individuals to be maintained in the population. This trick, however, may not be effective, depending on the landscape of the problem. Another possible technique would be to simply replace part of the population with randomly generated individuals, when most of the population is too similar to each other. Diversity is important in genetic algorithms (and genetic programming) because crossing over a homogeneous population does not yield new solutions. In evolution strategies and evolutionary programming, diversity is not essential because of a greater reliance on mutation.

- Operating on dynamic data sets is difficult, as genomes begin to converge early on towards solutions which may no longer be valid for later data. Several methods have been proposed to remedy this by increasing genetic diversity somehow and preventing early convergence, either by increasing the probability of mutation when the solution quality drops (called *triggered hypermutation*), or by occasionally introducing entirely new, randomly generated elements into the gene pool (called *random immigrants*). Again, evolution strategies and evolutionary programming can be implemented with a so-called "comma strategy" in which parents are not maintained and new parents are selected only from offspring. This can be more effective on dynamic problems.
- GAs cannot effectively solve problems in which the only fitness measure is a single right/wrong measure (like decision problems), as there is no way to converge on the solution (no hill to climb). In these cases, a random search may find a solution as quickly as a GA. However, if the situation allows the success/failure trial to be repeated giving (possibly) different results, then the ratio of successes to failures provides a suitable fitness measure.

- For specific optimisation problems and problem instances, other optimisation algorithms may find better solutions than genetic algorithms (given the same amount of computation time). Alternative and complementary algorithms include evolution strategies, evolutionary programming, simulated annealing, Gaussian adaptation, hill climbing, and swarm intelligence (e.g.: ant colony optimisation, particle swarm optimisation) and methods based on integer linear programming. The question of which, if any, problems are suited to genetic algorithms (in the sense that such algorithms are better than others) is open and controversial.

Variants

The simplest algorithm represents each chromosome as a bit string. Typically, numeric parameters can be represented by integers, though it is possible to use floating point representations. The floating point representation is natural to evolution strategies and evolutionary programming. The notion of real-valued genetic algorithms has been offered but is really a misnomer because it does not really represent the building block theory that was proposed by John Henry Holland in the 1970s. This theory is not without support though, based on theoretical and experimental results. The basic algorithm performs crossover and mutation at the bit level. Other variants treat the chromosome as a list of numbers which are indexes into an instruction table, nodes in a linked list, hashes, objects, or any other imaginable data structure. Crossover and mutation are performed so as to respect data element boundaries. For most data types, specific variation operators can be designed. Different chromosomal data types seem to work better or worse for different specific problem domains.

When bit-string representations of integers are used, Gray coding is often employed. In this way, small changes in the integer can be readily effected through mutations or crossovers. This has been found to help prevent premature convergence at so called *Hamming walls*, in which too many simultaneous mutations (or crossover events) must occur in order to change the chromosome to a better solution.

Other approaches involve using arrays of real-valued numbers instead of bit strings to represent chromosomes. Theoretically, the smaller the alphabet, the better the performance, but paradoxically, good results have been obtained from using real-valued chromosomes.

A very successful (slight) variant of the general process of constructing a new population is to allow some of the better organisms from the current generation to carry over to the next, unaltered. This strategy is known as *elitist selection.* Parallel implementations of genetic algorithms come in two flavours. Coarse-grained parallel genetic algorithms assume a population on each of the computer nodes and migration of individuals among the nodes. Fine-grained parallel genetic algorithms assume an individual on each processor node which acts with neighbouring individuals for selection and reproduction. Other variants, like genetic algorithms for online optimisation problems, introduce time-dependence or noise in the fitness function.

Genetic algorithms with adaptive parameters (adaptive genetic algorithms, AGAs) is another significant and promising variant of genetic algorithms. The probabilities of crossover (pc) and mutation (pm) greatly determine the degree of solution accuracy and the convergence speed that genetic algorithms can obtain. Instead of using fixed values of pc and pm, AGAs utilise the population information in each generation and adaptively adjust the pc and pm in order to maintain the population diversity as well as to sustain the convergence capacity. In AGA (adaptive genetic algorithm), the adjustment of pc and pm depends on the fitness values of the solutions. In *CAGA* (clustering-based adaptive genetic algorithm), through the use of clustering analysis to judge the optimisation states of the population, the adjustment of pc and pm depends on these optimisation states. The *GEGA* program is an *ab initio* gradient embedded GA, a program for finding the global minima of clusters developed by Anastassia Alexandrova at Utah State University. GEGA employs geometry-cuts for the GA, *ab initio* level of computation for geometry optimisation and vibrational frequency analysis, with local minima only, and a specific mutational procedure based on the so called "kick technique".

It can be quite effective to combine GA with other optimisation methods. GA tends to be quite good at finding generally good global solutions, but quite inefficient at finding the last few mutations to find the absolute optimum. Other techniques (such as simple hill climbing) are quite efficient at finding absolute optimum in a limited region. Alternating GA and hill climbing can improve the efficiency of GA while overcoming the lack of robustness of hill climbing.

This means that the rules of genetic variation may have a different meaning in the natural case. For instance – provided that steps are stored in consecutive order – crossing over may sum a number of steps from maternal DNA adding a number of steps from paternal DNA and so on. This is like adding vectors that more probably may follow a ridge in the phenotypic landscape. Thus, the efficiency of the process may be increased by many orders of magnitude. Moreover, the inversion operator has the opportunity to place steps in consecutive order or any other suitable order in favour of survival or efficiency.

A variation, where the population as a whole is evolved rather than its individual members, is known as gene pool recombination.

Linkage-learning

A number of variations have been developed to attempt to improve performance of GAs on problems with a high degree of fitness epistasis, i.e. where the fitness of a solution consists of interacting subsets of its variables. Such algorithms aim to learn (before exploiting) these beneficial phenotypic interactions. As such, they are aligned with the Building Block Hypothesis in adaptively reducing disruptive recombination. Prominent examples of this approach include the mGA, GEMGA and LLGA.

Problem Domains

Problems which appear to be particularly appropriate for solution by genetic algorithms include timetabling and scheduling problems, and many scheduling software packages are based on GAs. GAs have also been applied to engineering. Genetic algorithms are often applied as an approach to solve global optimisation problems.

As a general rule of thumb genetic algorithms might be useful in problem domains that have a complex fitness landscape as mixing, i.e., mutation in combination with crossover, is designed to move the population away from local optima that a traditional hill climbing algorithm might get stuck in. Observe that commonly used crossover operators cannot change any uniform population. Mutation alone can provide ergodicity of the overall genetic algorithm process.

Examples of problems solved by genetic algorithms include: mirrors designed to funnel sunlight to a solar collector, antennae designed to pick up radio signals in space, and walking methods for computer figures. Many of their solutions have been highly effective, unlike anything a human engineer would have produced, and inscrutable as to how they arrived at that solution.

History

Computer simulations of evolution started as early as in 1954 with the work of Nils Aall Barricelli, who was using the computer at the Institute for Advanced Study in Princeton, New Jersey. His 1954 publication was not widely noticed. Starting in 1957, the Australian quantitative geneticist Alex Fraser published a series of papers on simulation of artificial selection of organisms with multiple loci controlling a measurable trait. From these beginnings, computer simulation of evolution by biologists became more common in the early 1960s, and the methods were described in books by Fraser and Burnell (1970) and Crosby (1973). Fraser's simulations included all of the essential elements of modern genetic algorithms. In addition, Hans-Joachim Bremermann published a series of papers in the 1960s that also adopted a population of solution to optimisation problems, undergoing recombination, mutation, and selection. Bremermann's research also included the elements of modern genetic algorithms. Other noteworthy early pioneers include Richard Friedberg, George Friedman, and Michael Conrad. Many early papers are reprinted by Fogel (1998).

Although Barricelli, in work he reported in 1963, had simulated the evolution of ability to play a simple game, artificial evolution became a widely recognised optimisation method as a result of the work of Ingo Rechenberg and Hans-Paul Schwefel in the 1960s and early 1970s – Rechenberg's group was able to solve complex engineering problems through evolution strategies. Another approach was the evolutionary programming technique of Lawrence J. Fogel, which was proposed for generating artificial intelligence. Evolutionary programming originally used finite state machines for predicting environments, and used variation and selection to optimise the predictive logics. Genetic algorithms in particular became popular through the work of John Holland in the early 1970s, and particularly his book *Adaptation in Natural and Artificial Systems* (1975). His work originated with studies of cellular automata, conducted by Holland and his students at the University of Michigan. Holland introduced a formalised framework for predicting the quality of the next generation, known as Holland's Schema Theorem. Research in GAs remained largely theoretical until the mid-1980s, when The First International Conference on Genetic Algorithms was held in Pittsburgh, Pennsylvania.

As academic interest grew, the dramatic increase in desktop computational power allowed for practical application of the new technique. In the late 1980s, General Electric started selling the world's first genetic algorithm product, a mainframe-based toolkit designed for industrial processes. In 1989, Axcelis, Inc. released Evolver, the world's first commercial GA product for desktop computers. The New York Times technology writer John Markoff wrote about Evolver in 1990.

Load Profile

In electrical engineering, a load profile is a graph of the variation in the electrical load versus time. A load profile will vary according to customer type (typical examples include residential, commercial and industrial), temperature and holiday seasons.

Power Generation

In the electricity generation sector, a load curve is a chart showing the amount of electricity customers use over a period of time. Generation companies use this information to plan how much power they will need to generate at any given time. A load duration curve is similar to a load curve. The information is the same but is presented in a different form. These curves are useful in the selection of generator units for supplying electricity.

Electricity Distribution

In an electricity distribution network, the load profile of electricity usage is important to the efficiency and reliability of power transmission. The power transformer or battery-to-grid are critical aspects of power distribution and sizing and modelling of batteries or transformers depends on the load profile. The factory specification of transformers for the optimisation of load losses versus no-load losses is dependent directly on the characteristics of the load profile that the transformer is expected to be subjected to. This includes such characteristics as average load factor, utilisation factor, and responsibility factor, which can all be calculated based on a given load profile. A transformer typically has a cyclic rating allowing for the variation in the load profile.

Retail Energy Markets

In retail energy markets, supplier obligations are settled on an hourly or subhourly basis. For most customers, consumption is measured on a monthly basis, based on meter reading schedules. Load profiles are used to convert the monthly consumption data into estimates of hourly or subhourly consumption in order to determine the supplier obligation. For each hour, these estimates are aggregated for all customers of an energy supplier, and the aggregate amount is used in market settlement calculations as the total demand that must be covered by the supplier.

Calculating a Load Profile

Load profiles can be determined by direct metering but on smaller devices such as distribution network transformers this is not routinely done. Instead a load profile can be inferred from customer billing or other data.

An example of a practical calculation used by utilities is using a transformer's maximum demand reading and taking into account the known number of each customer type supplied by these transformers. This process is called load research.

Recording Load Profiles

Actual demand can be collected at strategic locations to perform more detailed load analysis; this is beneficial to both distribution and end-user customers looking for peak consumption. Smart grid meters, utility meter load profilers, data logging sub-meters and portable data loggers are designed to accomplish this task by recording readings at a set interval.

Low Voltage Ride Through

In electricity supply and generation, low voltage ride through (LVRT), or fault ride through (FRT), is what an electric device, especially wind generator, may be required to be capable of when the voltage in the grid is temporarily reduced due to a fault or load change in the grid. The voltage may be reduced in one, two or all the three phases of the AC grid. The severity of the voltage dip is defined by the voltage level during the dip (may go down to zero) and the duration of the dip.

Depending on the application the device may, during and after the dip, be required to:

- disconnect temporarily from the grid, but reconnect and continue operation after the dip
- stay operational and not disconnect from the grid
- stay connected and support the grid with reactive power (defined as the reactive current of the positive sequence of the fundamental)

For generating units such as wind turbines and solar power stations, the required LVRT behaviour is defined in grid codes issued by the grid operator. Examples of the such grid codes are the German E. On grid code and UK National Grid code.

For critical loads such as power supplies for computer centres and industrial processes the required LVRT performance is defined by the application. Often a special device called uninterruptible power supply (UPS) is used to provide power to the critical load during the voltage dip.

Chapter 5

Electric Power Distribution

Electricity distribution is the final stage in the delivery of electricity to end users. A distribution system's network carries electricity from the transmission system and delivers it to consumers. Typically, the network would include medium-voltage (less than 50 kV) power lines, substations and pole-mounted transformers, low-voltage (less than 1 kV) distribution wiring and sometimes meters.

History

In the early days of electricity distribution, direct current (DC) generators were connected to loads at the same voltage. The generation, transmission and loads had to be of the same voltage because there was no way of changing DC voltage levels, other than inefficient motor-generator sets. Low DC voltages were used (on the order of 100 volts) since that was a practical voltage for incandescent lamps, which were the primary electrical load. Low voltage also required less insulation for safe distribution within buildings. The losses in a cable are proportional to the square of the current, the length of the cable, and the resistivity of the material, and are inversely proportional to cross-sectional area. Early transmission networks used copper cable, which is one of the best economically feasible conductors for this application. To reduce the current and copper required for a given quantity of power transmitted would require a higher transmission voltage, but no efficient method existed to change the voltage of DC power circuits. To keep losses to an economically practical level the Edison DC system needed thick cables and local generators. Early DC generating plants needed to be within about 1.5 miles (2.4 km) of the farthest customer to avoid excessively large and expensive conductors.

Introduction of Alternating Current

The competition between the direct current (DC) of Thomas Edison and the alternating current (AC) of Nikola Tesla and George

Westinghouse was known as the War of Currents. At the conclusion of their campaigning, AC became the dominant form of transmission of power.

Power transformers, installed at power stations, could be used to raise the voltage from the generators, and transformers at local substations could reduce voltage to supply loads. Increasing the voltage reduced the current in the transmission and distribution lines and hence the size of conductors and distribution losses. This made it more economical to distribute power over long distances. Generators (such as hydroelectric sites) could be located far from the loads.

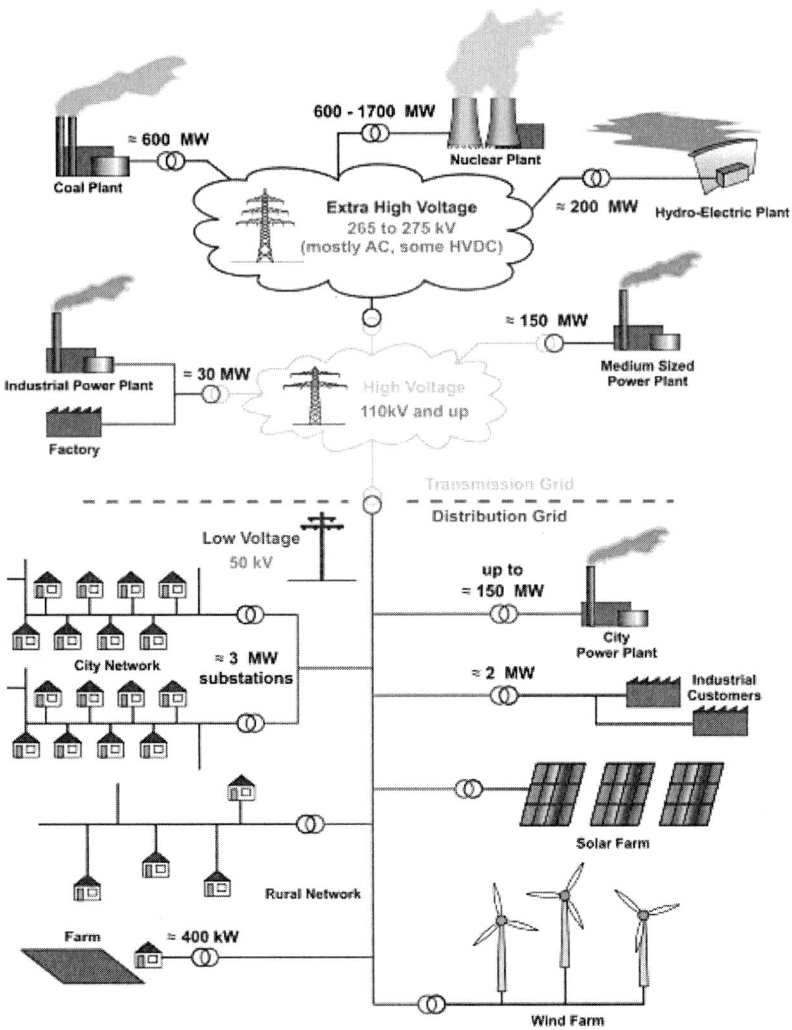

Figure : General layout of electricity networks

In North America, early distribution systems used a voltage of 2.2 kV corner-grounded delta. Over time, this was gradually increased to 2.4 kV. As cities grew, most 2.4 kV systems were upgraded to 2.4/4.16 kV, three-phase systems. In three phase networks that permit connections between phase and neutral, both the phase-to-phase voltage (4160, in this example) and the phase-to-neutral voltage are given, the network does not serve single-phase loads connected phase-to-neutral. Some city and suburban distribution systems continue to use this range of voltages, but most have been converted to 7200/12470Y, 7620/13200Y, 14400/24940Y, and 19920/34500Y.

European systems used 3.3 kV to ground, in support of the 220/380Y volt power systems used in those countries. In the UK, urban systems progressed to 6.6 kV and then 11 kV (phase to phase), the most common distribution voltage.

North American and European power distribution systems also differ in that North American systems tend to have a greater number of low-voltage step-down transformers located close to customers' premises. For example, in the US a pole-mounted transformer in a suburban setting may supply 7-8 houses, whereas in the UK a typical urban or suburban low-voltage substation would normally be rated between 315 kVA and 1 MVA and supply a whole neighbourhood. This is because the higher voltage used in Europe (415 V vs 230 V) may be carried over a greater distance with acceptable power loss. An advantage of the North American setup is that failure or maintenance on a single transformer will only affect a few customers. Advantages of the UK setup are that the transformers may be fewer, larger and more efficient, and due to diversity there need be less spare capacity in the transformers, reducing power wastage. In North American city areas with many customers per unit area, network distribution will be used, with multiple transformers and low-voltage buses interconnected over several city blocks.

Rural Electrification systems, in contrast to urban systems, tend to use higher voltages because of the longer distances covered by those distribution lines. 7.2, 12.47, 25, and 34.5 kV distribution is common in the United States; 11 kV and 33 kV are common in the UK, New Zealand and Australia; 11 kV and 22 kV are common in South Africa. Other voltages are occasionally used.

In New Zealand, Australia, Saskatchewan, Canada, and South Africa, single wire earth return systems (SWER) are used to electrify remote rural areas.

While power electronics now allow for conversion between DC voltage levels, AC is still used in distribution due to the economy, efficiency and reliability of transformers. High-voltage DC is used for transmission of large blocks of power over long distances, or for interconnecting adjacent AC networks, but not for distribution to customers. Electric power is normally generated at 11-25kV in a power station. To transmit over long distances, it is then stepped-up to 400kV, 220kV or 132kV as necessary. Power is carried through a transmission network of high voltage lines. Usually, these lines run into hundreds of kilometres and deliver the power into a common power pool called the grid. The grid is connected to load centres (cities) through a sub-transmission network of normally 33kV (or sometimes 66kV) lines. These lines terminate into a 33kV (or 66kV) substation, where the voltage is stepped-down to 11kV for power distribution to load points through a distribution network of lines at 11kV and lower.

Modern Distribution Systems

The modern distribution system begins as the primary circuit leaves the sub-station and ends as the secondary service enters the customer's meter socket. Distribution circuits serve many customers. The voltage used is appropriate for the shorter distance and varies from 2,300 to about 35,000 volts depending on utility standard practice, distance, and load to be served. Distribution circuits are fed from a transformer located in an electrical substation, where the voltage is reduced from the high values used for power transmission.

Conductors for distribution may be carried on overhead pole lines, or in densely-populated areas where they are buried underground. Urban and suburban distribution is done with three-phase systems to serve both residential, commercial, and industrial loads. Distribution in rural areas may be only single-phase if it is not economical to install three-phase power for relatively few and small customers.

Only large consumers are fed directly from distribution voltages; most utility customers are connected to a transformer, which reduces the distribution voltage to the relatively low voltage used by lighting and interior wiring systems. The transformer may be pole-mounted or set on the ground in a protective enclosure. In rural areas a pole-mount transformer may serve only one customer, but in more built-up areas multiple customers may be connected. In very dense city areas, a secondary network may be formed with many transformers feeding into a common bus at the utilisation voltage. Each customer has an "electrical service" or "service drop" connection and a meter for billing. (Some very small loads, such as yard lights, may be too small to meter and so are charged only a monthly rate.) A ground connection to local earth is normally provided for the customer's system as well as for the equipment owned by the utility.

The purpose of connecting the customer's system to ground is to limit the voltage that may develop if high voltage conductors fall on the lower-voltage conductors, or if a failure occurs within a distribution transformer. If all conductive objects are bonded to the same earth grounding system, the risk of electric shock is minimised. However, multiple connections between the utility ground and customer ground can lead to stray voltage problems; customer piping, swimming pools or other equipment may develop objectionable voltages. These problems may be difficult to resolve since they often originate from places other than the customer's premises.

International Differences

In many areas, "delta" three phase service is common. Delta service has no distributed neutral wire and is therefore less expensive. In North America and Latin America, three phase service is often a *Y* (*wye*) in which the neutral is directly connected to the center of the generator rotor.

The neutral provides a low-resistance metallic return to the distribution transformer. Wye service is recognisable when a line has four conductors, one of which is lightly insulated. Three-phase wye service is excellent for motors and heavy power use. Many areas in the world use single-phase 220 V or 230 V residential and light industrial service. In this system, the high voltage distribution network supplies a few substations per area, and the 230 V power from each substation is directly distributed. A live (hot) wire and neutral are connected to the building from one phase of three phase service. Single-phase distribution is used where motor loads are small.

Americas

In the U.S. and parts of Canada and Latin America, split phase service is the most common. Split phase provides both 120 V and 240 V service with only three wires. The house voltages are provided by local transformers.

The neutral is directly connected to the three-phase neutral. Socket voltages are only 120 V, but 240 V is available for heavy appliances because the two halves of a phase oppose each other.

Europe

In Europe, electricity is normally distributed for industry and domestic use by the three-phase, four wire system. This gives a three-phase voltage of 400 volts and a single-phase voltage of 230 volts. For industrial customers, 3-phase 690 / 400 volt is also available.

Japan

Japan has a large number of small industrial manufacturers, and therefore supplies standard low-voltage three phase-service in many suburbs. Also, Japan normally supplies residential service as two phases of a three phase service, with a neutral. These work well for both lighting and motors.

Rural Services

Rural services normally try to minimise the number of poles and wires. Single-wire earth return (SWER) is the least expensive, with one wire. It uses high voltages, which in turn permit use of galvanized steel wire. The strong steel wire permits inexpensive wide pole spacings. Other areas use high voltage split-phase or three phase service at higher cost.

Metering

Electricity meters use different *metering equations* depending on the form of electrical service. Since the math differs from service to service, the number of conductors and sensors in the meters also vary.

Terms

Besides referring to the physical wiring, the term *electrical service* also refers in an abstract sense to the provision of electricity to a building.

Distribution Network Configurations

Figure : Substation near Yellowknife, in the Northwest Territories of Canada

Distribution networks are typically of two types, radial or interconnected. A radial network leaves the station and passes through the network area with no normal connection to any other supply. This is typical of long rural lines with isolated load areas. An interconnected network is generally found in more urban areas and will have multiple connections to other points of supply. These points of connection are normally open but allow various configurations by the operating utility by closing and opening switches. Operation of these switches may be by remote control from a control center or by a lineman. The benefit of the interconnected model is that in the event of a fault or required maintenance a small area of network can be isolated and the remainder kept on supply.

Within these networks there may be a mix of overhead line construction utilising traditional utility poles and wires and, increasingly, underground construction with cables and indoor or cabinet substations. However, underground distribution is significantly more expensive than overhead construction. In part to reduce this cost, underground power lines are sometimes co-located with other utility lines in what are called common utility ducts. Distribution feeders emanating from a substation are generally controlled by a circuit breaker which will open when a fault is detected. Automatic circuit reclosers may be installed to further segregate the feeder thus minimising the impact of faults.

Long feeders experience voltage drop requiring capacitors or voltage regulators to be installed.

Characteristics of the supply given to customers are generally mandated by contract between the supplier and customer. Variables of the supply include:

- AC or DC - Virtually all public electricity supplies are AC today. Users of large amounts of DC power such as some electric railways, telephone exchanges and industrial processes such as aluminium smelting usually either operate their own or have adjacent dedicated generating equipment, or use rectifiers to derive DC from the public AC supply
- Voltage, including tolerance (usually +10 or -15 percent)
- Frequency, commonly 50 or 60 Hz, 16.6 Hz for some railways and, in a few older industrial and mining locations, 25 Hz.
- Phase configuration (single phase, polyphase including two-phase and three phase)

- Maximum demand (usually measured as the largest amount of power delivered within a 15 or 30 minute period during a billing period)
- Load factor, expressed as a ratio of average load to peak load over a period of time. Load factor indicates the degree of effective utilisation of equipment (and capital investment) of distribution line or system.
- Power factor of connected load
- Earthing arrangements - TT, TN-S, TN-C-S or TN-C
- Prospective short circuit current
- Maximum level and frequency of occurrence of transients.

Distribution Industry

Traditionally the electricity industry has been a publicly owned institution but starting in the 1970s nations began the process of deregulation and privatisation, leading to electricity markets. A major focus of these was the elimination of the former so called *natural monopoly* of generation, transmission, and distribution. As a consequence, electricity has become more of a commodity. The separation has also led to the development of new terminology to describe the business units (e.g., line company, wires business and network company).

AC Power Flow Model

AC power flow model is a model used in electrical engineering to analyse power grids. It provides a nonlinear system which describes the energy flow through each transmission line. Due to nonlinearity, in many cases the analysis of large network via AC power flow model is not feasible, and a linear (but less accurate) DC power flow model is used instead. Both of those models are very crude approximations to reality.

Advanced Configuration and Power Interface

In computing, the Advanced Configuration and Power Interface (ACPI) specification provides an open standard for device configuration and power management by the operating system.

First released in December 1996, *ACPI* defines platform-independent interfaces for hardware discovery, configuration, power management and monitoring. The specification is central to *Operating System-directed configuration and Power Management* (OSPM), a term used to describe a system implementing ACPI, which removes device management responsibilities from legacy firmware interfaces.

Overview

ACPI aims to consolidate, check and improve upon existing power and configuration standards for hardware devices. It provides a transition from earlier standards to entirely ACPI-compliant hardware, with some ACPI operating systems already removing support for legacy hardware. With the intention of replacing Advanced Power Management, the MultiProcessor Specification and the Plug and Play BIOS Specification, the standard brings power management under the control of the operating system (OSPM), as opposed to the previous BIOS-central system, which relied on platform-specific firmware to determine power management and configuration policy.

The ACPI specification contains numerous related components for hardware and software programming, as well as a unified standard for device/power interaction and for bus configuration. Some software developers have trouble implementing ACPI, and express concerns about the requirements that bytecode from an external source must be run by the system with full privileges. Linus Torvalds, creator of the Linux kernel, once described it as "a complete design disaster in every way" in relation to his view that "modern PCs are horrible".

Microsoft's Windows 98 became the first operating system with full support for ACPI, with Windows 2000, Windows XP, Windows Vista, Windows 7, eComStation, FreeBSD, NetBSD, OpenBSD, HP-UX, OpenVMS, Linux, and PC versions of SunOS all having at least some support for ACPI.

OSPM Responsibilities

ACPI requires that once an OSPM-compatible operating system has activated ACPI on a computer, it then takes over and has exclusive control of all aspects of power management and device configuration. The OSPM implementation must expose an ACPI-compatible environment to device drivers, which exposes certain system, device and processor states.

Power States

Global States

The ACPI specification defines the following four Global "Gx" states and six Sleep "Sx" states for an ACPI-compliant computer-system:

- G0 (S0): *Working*. "Awaymode" is a subset of S0, where monitor is off but background tasks are running.
- G1, *Sleeping* subdivides into the four states S1 through S4:
  - o S1: All processor caches are flushed, and the CPU(s) stop executing instructions. Power to the CPU(s) and RAM is maintained; devices that do not indicate they must remain on may be powered down.
  - o S2: CPU powered off. Dirty cache is flushed to RAM.
  - o S3: Commonly referred to as *Standby*, *Sleep*, or *Suspend to RAM*. RAM remains powered
  - o S4: *Hibernation* or *Suspend to Disk*. All content of main memory is saved to non-volatile memory such as a hard drive, and is powered down.
- G2 (S5), *Soft Off*: G2/S5 is almost the same as G3 *Mechanical Off*, except that the PSU still supplies power, at a minimum, to the power button to allow return to S0. A full reboot is required. No previous content is retained. Other components may remain powered so the computer can "wake" on input from the keyboard, clock, modem, LAN, or USB device.
- G3, *Mechanical Off*: The computer's power has been totally removed via a mechanical switch (as on the rear of a PSU). The power cord can be removed and the system is safe for disassembly (typically, only the real-time clock continues to run-using its own small battery).

Furthermore, the specification defines a *Legacy* state: the state on an operating system which does not support ACPI. In this state, the hardware and power are not managed via ACPI, effectively disabling ACPI.

Device States

The device states *D0-D3* are device-dependent:

- D0 *Fully On* is the operating state.
- D1 and D2 are intermediate power-states whose definition varies by device.
- D3 *Off* has the device powered off and unresponsive to its bus.

Processor States

The CPU power states *C0-C3* are defined as follows:

- C0 is the operating state.

- C1 (often known as *Halt*) is a state where the processor is not executing instructions, but can return to an executing state essentially instantaneously. All ACPI-conformant processors must support this power state. Some processors, such as the Pentium 4, also support an Enhanced C1 state (C1E or Enhanced Halt State) for lower power consumption.
- C2 (often known as *Stop-Clock*) is a state where the processor maintains all software-visible state, but may take longer to wake up. This processor state is optional.
- C3 (often known as *Sleep*) is a state where the processor does not need to keep its cache coherent, but maintains other state. Some processors have variations on the C3 state (Deep Sleep, Deeper Sleep, etc.) that differ in how long it takes to wake the processor. This processor state is optional.

Performance States

While a device or processor operates (D0 and C0, respectively), it can be in one of several power-performance states. These states are implementation-dependent. Though, P0 is always the highest-performance state; with P1 to Pn being successively lower-performance states, up to an implementation-specific limit of n no greater than 16.

P-states have become known as SpeedStep in Intel processors, as PowerNow! or Cool'n'Quiet in AMD processors, and as PowerSaver in VIA processors.

- P0 max power and frequency
- P1 less than P0, voltage/frequency scaled
- ...
- Pn less than *P(n-1)*, voltage/frequency scaled.

Hardware Interface

ACPI-compliant systems interact with hardware through either a "Function Fixed Hardware (FFH) Interface", or a platform-independent hardware programming model which relies on platform-specific ACPI Machine Language (AML) provided by the original equipment manufacturer (OEM).

Function Fixed Hardware interfaces are platform-specific features, provided by platform manufacturers for the purposes of performance and failure recovery. Standard Intel-based PCs have a fixed function interface defined by Intel, which provides a set of core functionality that reduces an ACPI-compliant system's need for full driver stacks for providing basic functionality during boot time or in the case of major system failure.

Firmware Interface

ACPI defines a large number of tables that provide the interface between an ACPI-compliant operating system, and system firmware. For example:

- DSDT – Differentiated System Description Table
- SSDT – Secondary System Description Table
- SRAT – System Resource Affinity Table.

ACPI Component Architecture (ACPICA)

The ACPI Component Architecture (ACPICA) provides an open-source OS-independent reference implementation of the ACPI specification.

History

The first revision of the ACPI specification was released in December 1996, supporting 16 and 32-bit addressing spaces. It wasn't until August 2000 that ACPI received 64-bit address support as well as support for multiprocessor workstations and servers with revision 2.0. In September 2004, revision 3.0 gave the ACPI specification support for SATA connectors, PCI Express bus, >256 multiprocessor support, ambient light sensors and user-presence devices, as well as extending the Thermal model beyond the previous processor-centric support. In June 2009, the 4.0 specification added many new features to the design; most notable are USB 3.0 support, logical processor idling support, and x2APIC support. The latest of the major publications is revision 5.0, released in November 2011.

Alternator Synchronisation

The process of connecting an AC generator (alternator) to other AC generators is known as synchronisation and is crucial for the generation of AC electrical power.

A DC generator can be connected to a power network by adjusting its open-circuit terminal voltage to match the network voltage by either adjusting its speed or its field excitation; the exact engine speed is not critical. However, an AC machine must match both the amplitude and the timing of the network voltage, which requires both speed and excitation to be systematically and closely controlled for synchronisation. This extra complexity was one of the arguments against AC operation during the War of Currents in the 1880's. In modern systems, synchronisation of generators is carried out by automatic systems.

Conditions

There are five conditions that must be met before the synchronisation process takes place. The alternator must have equal *line voltage*, *frequency*, *phase sequence*, *phase angle* and *waveform* to that of the system to which it is being synchronised. Waveform and phase sequence are fixed by the construction of the generator and its connections to the system, but voltage, frequency and phase angle must be controlled each time a generator is to be connected to a grid.

Process

In the past, synchronisation was performed manually using three-lamp method. Nowadays, the process is automatically operated and controlled with the aid of synchronisation relays.

During installation of a generator, careful checks are made to ensure the generator terminals and all control wiring are correct so that the order of phases (phase sequence) matches the system. Connecting a generator with the wrong phase sequence will result in a short circuit as the system voltages are opposite to those of the generator terminal voltages.

The sequence of events is similar for manual or automatic synchronisation. The generator is brought up to approximate synchronous speed by supplying more energy to its shaft-for example, opening the valves on a steam turbine, opening the gates on a hydraulic turbine, or increasing the fuel rack setting on a diesel engine. The field of the generator is energised and the voltage at the terminals of the generator is observed and compared with the system. The voltage magnitude must be the same as the system voltage.

Synchronising Lamps

Formerly, three light bulbs were connected between the generator terminals and the system terminals (or more generally, to the terminals of instrument transformers connected to generator and system). As the generator speed changes, the lights will rise and fall in intensity at a rate proportional to the difference between generator frequency and system frequency. When the voltage at the generator is opposite to the system voltage (either ahead or behind in phase), the lamps will be bright. When the voltage at the generator matches the system voltage, the lights will be dark. At that instant, the circuit breaker connecting the generator to the system may be closed and the generator will then stay in synchronism with the system.

Synchroscope

Figure : From top to bottom: synchroscope, voltmeter, frequency meter. When the two systems are synchoronised, the pointer on the synchrosope is stationary and points straight up.

Another manual method of synchronisation relies on observing an instrument called a "synchroscope", which displays the relative frequencies of system and generator.

The pointer of the synchroscope will indicate "fast" or "slow" speed of the generator with respect to the system. To minimise the transient current when the generator circuit breaker is closed, usual practice is to initiate the close as the needle slowly approaches the in-phase point. An error of a few electrical degrees between system and generator will result in a momentary inrush and abrupt speed change of the generator.

Synchronising Relays

Synchronising relays allow unattended synchronisation of a machine with a system. Today these are digital microprocessor instruments, but in the past electromechanical relay systems were applied. A synchronising relay is useful to remove human reaction time from the process, or when a human is not available such as at a remote controlled generating plant. Sometimes as a precaution against out-of-step connection of a machine to a system, a "synchro check" relay is installed that prevents closing the generator circuit breaker unless the machine is within a few electrical degrees of being in-phase with the system. (Synchro check relays are also applied in places where several sources of supply may be connected and where it is important that out-of-step sources are not accidentally paralleled.)

Synchronous operation : When the generator is synchronised, the frequency of the system will change depending on load and the average characteristics of all the generating units connected to the grid. Large changes in system frequency can cause the generator to fall out of synchronism with the system. Protective devices on the generator will operate to disconnect it automatically.

Amplitude Adjusting

The Amplitude adjusting (also referred to as Amplitude control) enables the power control of electric loads, which are operated with AC voltage. A representative application is the heating control of industrial high temperature furnaces.

Functionality

Contrary to the conventional phase angle or full wave control, during amplitude control only the Amplitude of the sinusoidal supply current is changed. The level of the amplitude only depends on the consumed power. The sinus oscillation does not change.

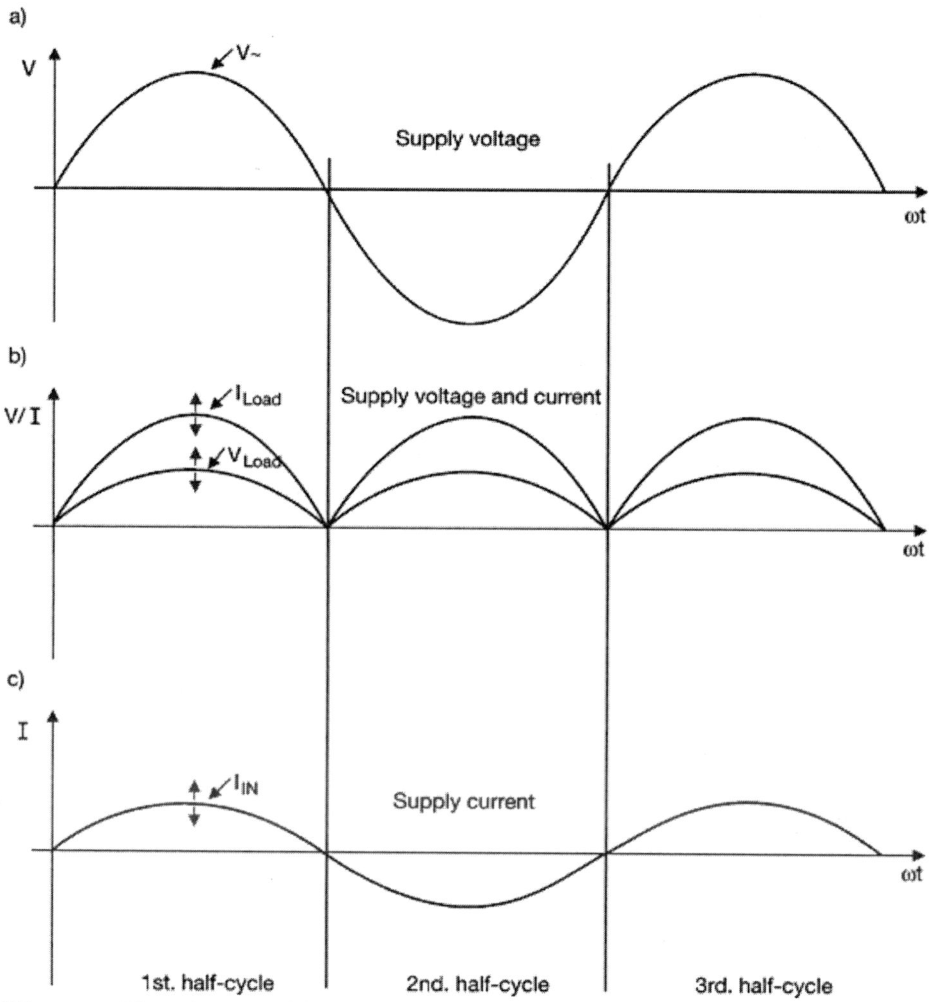

Figure : The sinus oscillation does not change.

Because current and voltage are in phase, only real power is taken from the mains for amplitude control. So the current consumption from the mains is considerably lower than the current consumption in case of phase-angle operation.

Advantages

The continuous current flow causes a mild operation of the used heater elements and consequently significant longer lifetimes are realised. Depending on the ambient conditions the lifetime can be twice as long. Especially the surface damage of the heater elements at thresholds can be reduced.

The amplitude control eliminates the flicker effects and harmonics, as usual for Thyristor units, so that the standard specifications according to EN 61000-3-2 and EN 61000-3-3 are observed.

Reactive power compensation is not required, reducing equipment costs.

Applications

Sinus units or IGBT power converters for power control of:

- Resistance heatings
- Silicon carbide (SC) - heater elements
- Molybdenum disilicide (MoSi2) - heater elements
- Infralight radiators.

Blondel's Theorem

Blondel's theorem, named after its discoverer, French electrical engineer André Blondel, states that, in a system of N electrical conductors, N-1 electrical meter or wattmeter elements, when properly connected, will measure the electrical power or energy taken. The connection must be such that all potential coils have a common tie to the conductor in which there is no current coil. For example, a 3 phase 3 wire system needs only two watt meters, but a 3 phase 4 wire system (with a neutral) needs three.

The theorem is based on a paper delivered by Blondel in 1893 at the International Electric Congress in Chicago.

Figure : A 50 Hz electrical substation in Melbourne. This is showing three of the five 220 kV/66 kV transformers, each with a capacity of 185 MVA

Figure : A 115 kV to 41.6/12.47 kV 5 MVA 60 Hz substation with circuit switcher, regulators, reclosers and control building at Warren, Minnesota

A substation is a part of an electrical generation, transmission, and distribution system. Substations transform voltage from high to low, or the reverse, or perform any of several other important functions. Electric power may flow through several substations between generating plant and consumer, and its voltage may change in several steps.

Substations may be owned and operated by a transmission or generation electrical utility, or may be owned by a large industrial or commercial customer. Generally substations are un-attended, relying on SCADA for remote supervision and control.

A substation may include transformers to change voltage levels between high transmission voltages and lower distribution voltages, or at the interconnection of two different transmission voltages. The word *substation* comes from the days before the distribution system became a grid. As central generation stations became larger, smaller generating plants were converted to distribution stations, receiving their energy supply from a larger plant instead of using their own generators. The first substations were connected to only one power station, where the generators were housed, and were subsidiaries of that power station.

Elements of a Substation

Substations generally have switching, protection and control equipment, and transformers. In a large substation, circuit breakers are used to interrupt any short circuits or overload currents that may occur on the network. Smaller distribution stations may use recloser circuit breakers or fuses for protection of distribution circuits. Substations themselves do not usually have generators, although a power plant may have a substation nearby. Other devices such as capacitors and voltage regulators may also be located at a substation.

Figure : SF_6 110 kV instrument current transformer TGFM series, Russia

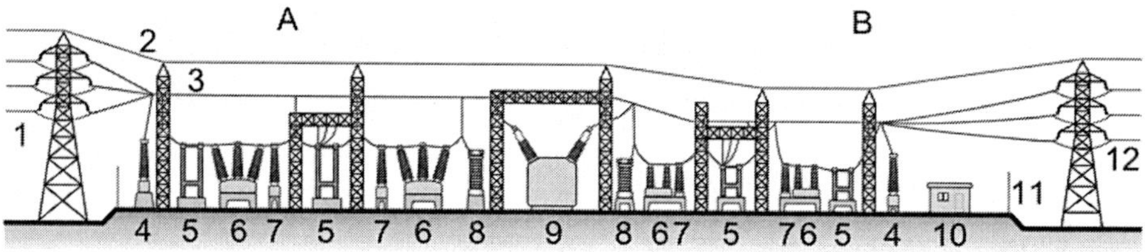

Figure : Elements of a substation

A:Primary power lines' side B:Secondary power lines' side :

1. Primary power lines 2. Ground wire 3. Overhead lines 4. Transformer for measurement of electric voltage 5. Disconnect switch 6. Circuit breaker 7. Current transformer 8. Lightning arrester 9. Main transformer 10. Control building 11. Security fence 12. Secondary power lines

Substations may be on the surface in fenced enclosures, underground, or located in special-purpose buildings. High-rise buildings may have several indoor substations. Indoor substations are usually found in urban areas to reduce the noise from the transformers, for reasons of appearance, or to protect switchgear from extreme climate or pollution conditions.

Where a substation has a metallic fence, it must be properly grounded (UK: earthed) to protect people from high voltages that may occur during a fault in the network. Earth faults at a substation can cause a ground potential rise. Currents flowing in the Earth's surface during a fault can cause metal objects to have a significantly different voltage than the ground under a person's feet; this *touch potential* presents a hazard of electrocution.

Transmission Substation

A transmission substation connects two or more transmission lines. The simplest case is where all transmission lines have the same voltage. In such cases, the substation contains high-voltage switches that allow lines to be connected or isolated for fault clearance or maintenance. A transmission station may have transformers to convert between two transmission voltages, voltage control/power factor correction devices such as capacitors, reactors or static VAr compensators and equipment such as phase shifting transformers to control power flow between two adjacent power systems. Transmission substations can range from simple to complex. A small "switching station" may be little more than a bus plus some circuit breakers. The largest transmission substations can cover a large area (several acres/ hectares) with multiple voltage levels, many circuit breakers and a large amount of protection and control equipment (voltage and current transformers, relays and SCADA systems). Modern substations may be implemented using international standards such as IEC Standard 61850.

Distribution Substation

A *distribution substation* transfers power from the transmission system to the distribution system of an area. It is uneconomical to directly connect electricity consumers to the main transmission network, unless they use large amounts of power, so the distribution station reduces voltage to a value suitable for local distribution.

The input for a distribution substation is typically at least two transmission or subtransmission lines. Input voltage may be, for example, 115 kV, or whatever is common in the area. The output is a number of feeders. Distribution voltages are typically medium voltage, between 2.4 kV and 33 kV depending on the size of the area served and the practices of the local utility.

The feeders run along streets overhead (or underground, in some cases) and power the distribution transformers at or near the customer premises. In addition to transforming voltage, distribution substations also isolate faults in either the transmission or distribution systems. Distribution substations are typically the points of voltage regulation, although on long distribution circuits (of several miles/kilometers), voltage regulation equipment may also be installed along the line.

The downtown areas of large cities feature complicated distribution substations, with high-voltage switching, and switching and backup systems on the low-voltage side. More typical distribution substations have a switch, one transformer, and minimal facilities on the low-voltage side.

Collector Substation

In distributed generation projects such as a wind farm, a collector substation may be required. It resembles a distribution substation although power flow is in the opposite direction, from many wind turbines up into the transmission grid. Usually for economy of construction the collector system operates around 35 kV, and the collector substation steps up voltage to a transmission voltage for the grid. The collector substation can also provide power factor correction if it is needed, metering and control of the wind farm. In some special cases a collector substation can also contain an HVDC static inverter plant.

Collector substations also exist where multiple thermal or hydroelectric power plants of comparable output power are in proximity. Examples for such substations are Brauweiler in Germany and Hradec in the Czech Republic, where power is collected from nearby lignite-fired power plants. If no transformers are required for increase of voltage to transmission level, the substation is a switching station.

Stations with Change of Current Type

Substations may be associated with HVDC converter plants, traction current, or interconnected non-synchronous networks. Formerly rotary converters changed frequency to interconnect two systems; such substations today are rare.

Switching Substation

A switching substation is a substation which does not contain transformers and operates only at a single voltage level. Switching substations are sometimes used as collector and distribution stations. Sometimes they are used for switching the current to back-up lines or for parallelising circuits in case of failure. An example is the switching stations for the HVDC Inga-Shaba transmission line.

Design

The main issues facing a power engineer are reliability and cost. A good design attempts to strike a balance between these two, to achieve sufficient reliability without excessive cost. The design should also allow expansion of the station, when required.

Selection of the location of a substation must consider many factors. Sufficient land area is required for installation of equipment with necessary clearances for electrical safety, and for access to maintain large apparatus such as transformers. Where land is costly, such as in urban areas, gas insulated switchgear may save money overall. The site must have room for expansion due to load growth or planned transmission additions. Environmental effects of the substation must be considered, such as drainage, noise and road traffic effects. A grounding (earthing) system must be designed, and ground potential rise must be calculated to protect passers-by during a short-circuit in the transmission system. The substation site must be reasonably central to the distribution area to be served.

The first step in planning a substation layout is the preparation of a one-line diagram which shows in simplified form the switching and protection arrangement required, as well as the incoming supply lines and outgoing feeders or transmission lines. It is a usual practice by many electrical utilities to prepare one-line diagrams with principal elements (lines, switches, circuit breakers, transformers) arranged on the page similarly to the way the apparatus would be laid out in the actual station.

In a common design, incoming lines have a disconnect switch and a circuit breaker. In some cases, the lines will not have both, with either a switch or a circuit breaker being all that is considered necessary. A disconnect switch is used to provide isolation, since it cannot interrupt load current. A circuit breaker is used as a protection device to interrupt fault currents automatically, and may be used to switch loads on and off, or to cut off a line when power is flowing in the 'wrong' direction. When a large fault current flows through the circuit breaker, this is detected through the use of current transformers. The magnitude of the current transformer outputs may be used to trip the circuit breaker resulting in a disconnection of the load supplied by the circuit break from the feeding point. This seeks to isolate the fault point from the rest of the system, and allow the rest of the system to continue operating with minimal impact. Both switches and circuit breakers may be operated locally (within the substation) or remotely from a supervisory control center.

Once past the switching components, the lines of a given voltage connect to one or more buses. These are sets of bus bars, usually in multiples of three, since three-phase electrical power distribution is largely universal around the world. The arrangement of switches, circuit breakers and buses used affects the cost and reliability of the substation. For important substations a ring bus, double bus, or so-called "breaker and a half" setup can be used, so that the failure of any one circuit breaker does not interrupt power to other circuits, and so that parts of the substation may be de-energised for maintenance and repairs. Substations feeding only a single industrial load may have minimal switching provisions, especially for small installations.

Once having established buses for the various voltage levels, transformers may be connected between the voltage levels. These will again have a circuit breaker, much like transmission lines, in case a transformer has a *fault* (commonly called a "short circuit").

Along with this, a substation always has control circuitry needed to command the various breakers to open in case of the failure of some component.

Switching Function

An important function performed by a substation is switching, which is the connecting and disconnecting of transmission lines or other components to and from the system. Switching events may be "planned" or "unplanned".

A transmission line or other component may need to be deenergised for maintenance or for new construction, for example, adding or removing a transmission line or a transformer.

To maintain reliability of supply, no company ever brings down its whole system for maintenance. All work to be performed, from routine testing to adding entirely new substations, must be done while keeping the whole system running.

Perhaps more important, a fault may develop in a transmission line or any other component. Some examples of this: a line is hit by lightning and develops an arc, or a tower is blown down by high wind. The function of the substation is to isolate the faulted portion of the system in the shortest possible time.

There are two main reasons: a fault tends to cause equipment damage; and it tends to destabilise the whole system. For example, a transmission line left in a faulted condition will eventually burn down; similarly, a transformer left in a faulted condition will eventually blow up. While these are happening, the power drain makes the system more unstable. Disconnecting the faulted component, quickly, tends to minimise both of these problems.

Automation

Early electrical substations required manual switching or adjustment of equipment, and manual collection of data for load, energy consumption, and abnormal events. As the complexity of distribution networks grew, it became economically necessary to automate supervision and control of substations from a centrally attended point, to allow overall coordination in case of emergencies and to reduce operating costs. Early efforts to remote control substations used dedicated communication wires, often run along side power circuits.

Power-line carrier, microwave radio, fibre optic cables as well as dedicated wired remote control circuits have all been applied to Supervisory Control and Data Acquisition (SCADA) for substations. The development of the microprocessor made for an exponential increase in the number of points that could be economically controlled and monitored.

Today, standardised communication protocols such as DNP3, IEC 61850 and Modbus, to list a few, are used to allow multiple intelligent electronic devices to communicate with each other and supervisory control centres. Distributed automatic control at substations is one element of the so-called smart grid.

Railways

Electrified railways also use substations, often distribution substations. In some cases a conversion of the current type takes place, commonly with rectifiers for direct current (DC) trains, or rotary converters for trains using alternating current (AC) other than that of the public grid. Sometimes they are also transmission substations or collector substations if the railway network also operates its own grid and generators.

Traction Substation

A traction substation or traction current converter plant is an electrical substation that converts electric power from the form provided by the electrical power industry for public utility service to an appropriate voltage, current type and frequency to supply railways, trams (streetcars) or trolleybuses with traction current.

Conversions

These systems can be used to convert three-phase 50Hz or 60Hz alternating current (AC) for the supply of AC railway electrification systems at a lower frequency and single phase, as used by many older systems, or to rectify AC into direct current (DC) for those systems (primarily public transit systems) using DC for traction power.

Equipment

Rotating

Originally, the conversion equipment usually consisted of one or more motor-generator sets containing three-phase synchronous AC motors and single-phase AC generators, mechanically coupled to a common shaft. Rotary converters were also used, especially where the desired output was DC current from an AC source.

Static

In the 1920s, DC was derived using electronic valves (mercury arc rectifiers). In modern systems, high-voltage DC (HVDC) "back-to-back" stations are used instead of mechanical equipment to convert between different frequencies and phases of AC power and solid-state thyristor rectifier systems are used for conversion from AC power to DC traction power.

Location

Traction current converter plants are either decentralised (where one plant directly supplies the overhead lines or third rail of the traction system, with no feed into a traction current distribution network) or centralised (for the supply of the traction power network, usually in addition to the direct supply of the overhead lines or third rail).

Central traction current converter plants are generally found in Germany (primarily in the cities of Neckarwestheim, Ulm, Nuremberg), Austria and Switzerland, while decentralised traction current converter plants are generally found in Norway, Sweden and the German states of Mecklenburg-Vorpommern and Brandenburg as well as parts of Great Britain. A list of current systems for electric rail traction provides further detail.

Chapter 6

Emergency Power System

Emergency power systems are a type of system, which may include lighting, generators, fuel cells and other apparatus, to provide backup power resources in a crisis or when regular systems fail. They find uses in a wide variety of settings from residential homes to hospitals, scientific laboratories, data centres, telecommunication equipment and modern naval ships. Emergency power systems can rely on generators, deep cycle batteries, flywheel energy storage or hydrogen fuel cells. Finally, some homebrew emergency power systems use regular lead-acid car batteries.

Figure : A backup power fuel cell for telecom applications

Flexible AC Transmission System

A flexible alternating current transmission system (FACTS) is a system composed of static equipment used for the AC transmission of electrical energy. It is meant to enhance controllability and increase power transfer capability of the network. It is generally a power electronics-based system.

Figure : A backup generator for a large apartment building

FACTS is defined by the IEEE as "a power electronic based system and other static equipment that provide control of one or more AC transmission system parameters to enhance controllability and increase power transfer capability."

Technology

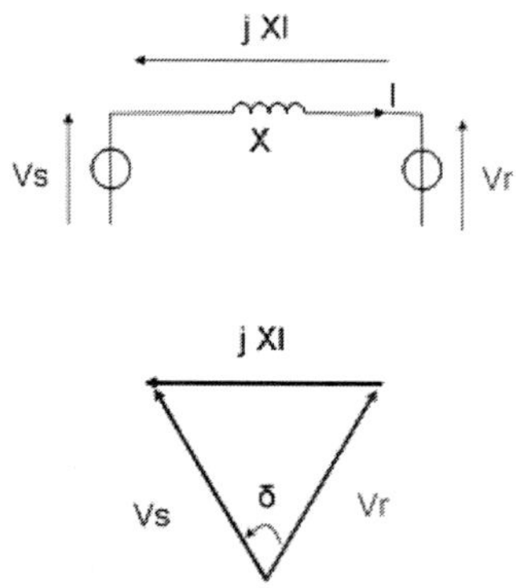

Figure : Transmission on a no-loss line.

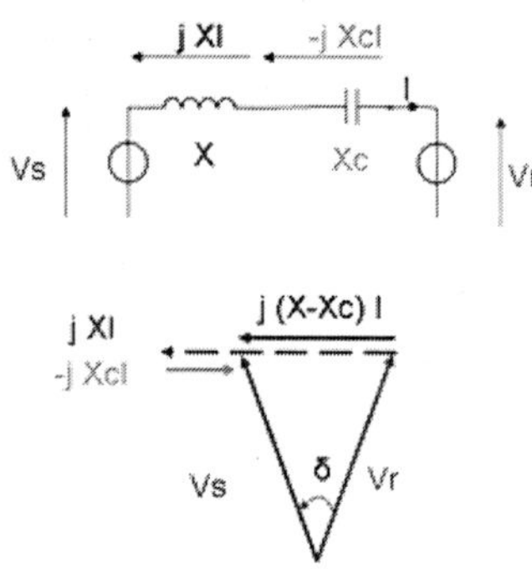

Figure : Series compensation.

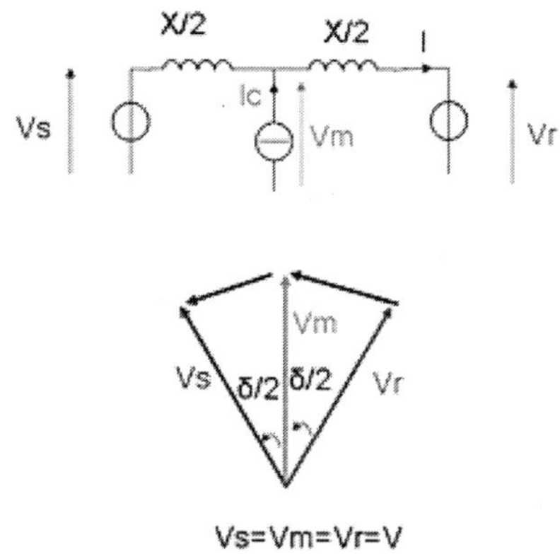

Figure : Shunt compensation.

History

The first FACTS installation was at the C. J. Slatt Substation in Northern Oregon. This is a 500 kV, 3-phase 60 Hz substation, and was developed by EPRI, the Bonneville Power Administration and General Electric Company.

Series Compensation

In series compensation, the FACTS is connected in series with the power system. It works as a controllable voltage source. Series inductance exists in all AC transmission lines. On long lines, when a large current flows, this causes a large voltage drop. To compensate, series capacitors are connected, decreasing the effect of the inductance.

Shunt Compensation

In shunt compensation, power system is connected in shunt (parallel) with the FACTS. It works as a controllable current source. Shunt compensation is of two types:

Shunt Capacitive Compensation : This method is used to improve the power factor. Whenever an inductive load is connected to the transmission line, power factor lags because of lagging load current. To compensate, a shunt capacitor is connected which draws current leading the source voltage. The net result is improvement in power factor.

Shunt Inductive Compensation : This method is used either when charging the transmission line, or, when there is very low load at the receiving end. Due to very low, or no load – very low current flows through the transmission line. Shunt capacitance in the transmission line causes voltage amplification (Ferranti Effect). The receiving end voltage may become double the sending end voltage (generally in case of very long transmission lines).

To compensate, shunt inductors are connected across the transmission line. The power transfer capability is thereby increased depending upon the power equation

$$P = \left(\frac{EV}{X}\right)\sin(\delta)$$

$\delta =$ power angle

Theory

In the case of a no-loss line, voltage magnitude at the receiving end is the same as voltage magnitude at the sending end:

$$V_s = V_r = V.$$

Transmission results in a phase lag δ that depends on line reactance X.

$$\underline{V_s} = V\cos\left(\frac{\delta}{2}\right) + jV\sin\left(\frac{\delta}{2}\right)$$

$$\underline{V_r} = V\cos\left(\frac{\delta}{2}\right) - jV\sin\left(\frac{\delta}{2}\right)$$

$$\underline{I} = \frac{\underline{V_s} - \underline{V_r}}{jX} = \frac{2V\sin\left(\frac{\delta}{2}\right)}{X}$$

As it is a no-loss line, active power P is the same at any point of the line:

$$P_s = P_r = P = V\cos\left(\frac{\delta}{2}\right)\cdot\frac{2V\sin\left(\frac{\delta}{2}\right)}{X} = \frac{V^2}{X}\sin(\delta)$$

Reactive power at sending end is the opposite of reactive power at receiving end:

$$Q_s = -Q_r = Q = V\sin\left(\frac{\delta}{2}\right)\cdot\frac{2V\sin\left(\frac{\delta}{2}\right)}{X} = \frac{V^2}{X}(1-\cos\delta)$$

As δ is very small, active power mainly depends on whereas reactive power mainly depends on voltage magnitude.

Series Compensation

FACTS for series compensation modify line impedance: X is decreased so as to increase the transmittable active power. However, more reactive power must be provided.

$$P = \frac{V^2}{X - Xc}\sin(\delta)$$

$$Q = \frac{V^2}{X - Xc}(1 - \cos\delta)$$

Shunt Compensation

Reactive current is injected into the line to maintain voltage magnitude. Transmittable active power is increased but more reactive power is to be provided.

$$P = \frac{2V^2}{X}\sin\left(\frac{\delta}{2}\right)$$

$$Q = \frac{2V^2}{X}\left[1 - \cos\left(\frac{\delta}{2}\right)\right]$$

Examples of Series Compensation

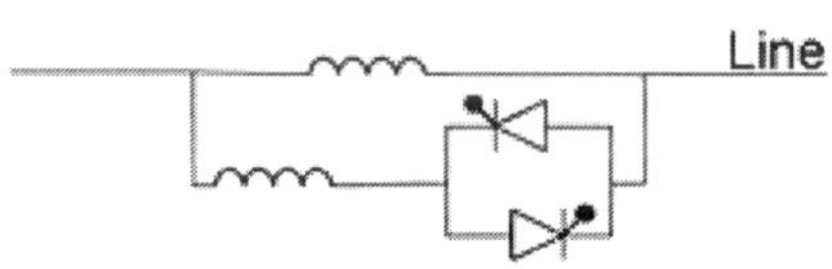

TCSR / TSSR

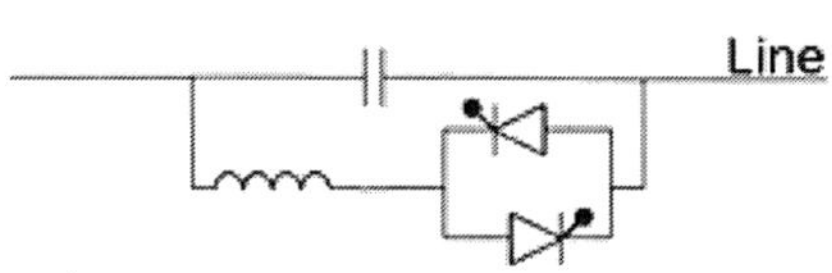

TCSC / TSSC

Examples of FACTS for series compensation (schematic):

- Static synchronous series compensator (SSSC)
- Thyristor-controlled series capacitor (TCSC): a series capacitor bank is shunted by a thyristor-controlled reactor
- Thyristor-controlled series reactor (TCSR): a series reactor bank is shunted by a thyristor-controlled reactor

- Thyristor-switched series capacitor (TSSC): a series capacitor bank is shunted by a thyristor-switched reactor
- Thyristor-switched series reactor (TSSR): a series reactor bank is shunted by a thyristor-switched reactor

Examples of Shunt Compensation

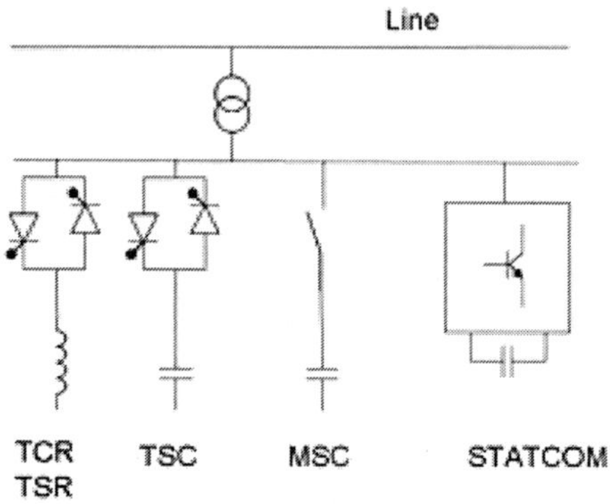

Examples of FACTS for shunt compensation (schematic):

- Static synchronous compensator (STATCOM); previously known as a static condenser (STATCON)
- Static VAR compensator (SVC). Most common SVCs are:
  - o Thyristor-controlled reactor (TCR): reactor is connected in series with a bidirectional thyristor valve. The thyristor valve is phase-controlled. Equivalent reactance is varied continuously.
  - o Thyristor-switched reactor (TSR): Same as TCR but thyristor is either in zero- or full- conduction. Equivalent reactance is varied in stepwise manner.
  - o Thyristor-switched capacitor (TSC): capacitor is connected in series with a bidirectional thyristor valve. Thyristor is either in zero- or full- conduction. Equivalent reactance is varied in stepwise manner.
  - o Mechanically-switched capacitor (MSC): capacitor is switched by circuit-breaker. It aims at compensating steady state reactive power. It is switched only a few times a day.

Grid-tied Electrical System

A grid-tied electrical system, also called *tied to grid* or *grid tie system*, is a semi-autonomous electrical generation or grid energy storage system which links to the mains to feed excess capacity back to the local mains electrical grid. When insufficient electricity is generated, or the batteries are not fully charged, electricity drawn from the mains grid can make up the shortfall.

When batteries are used to store, the system is called battery-to-grid (B2G), which includes vehicle-to-grid (V2G).

Grid-tie Inverter

A grid-tie inverter (GTI) is a special type of inverter that converts direct current(DC) electricity into alternating current(AC) electricity and feeds it into an existing electrical grid. GTIs are often used to convert direct current produced by many renewable energy sources, such as solar panels or small wind turbines, into the alternating current used to power homes and businesses. The technical name for a grid-tie inverter is "grid-interactive inverter". They may also be called synchronous inverters. Grid-interactive inverters typically cannot be used in standalone applications where utility power is not available.

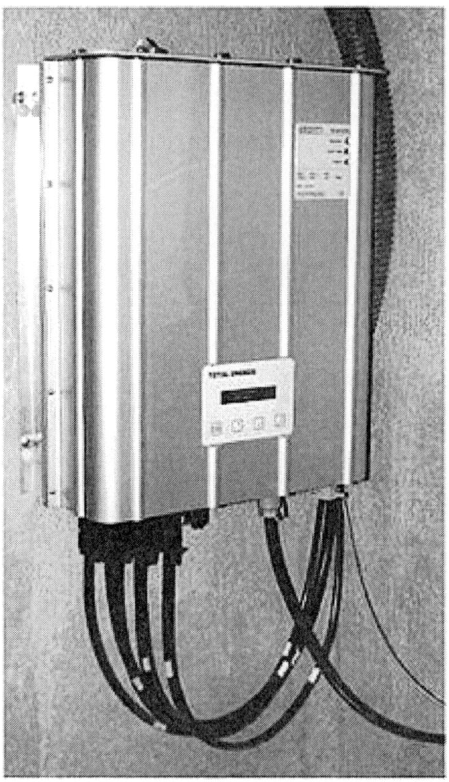

Figure : Inverter for grid connected PV

Figure : Example of large 3-phase inverter for commercial and utility scale grid-tied PV systems

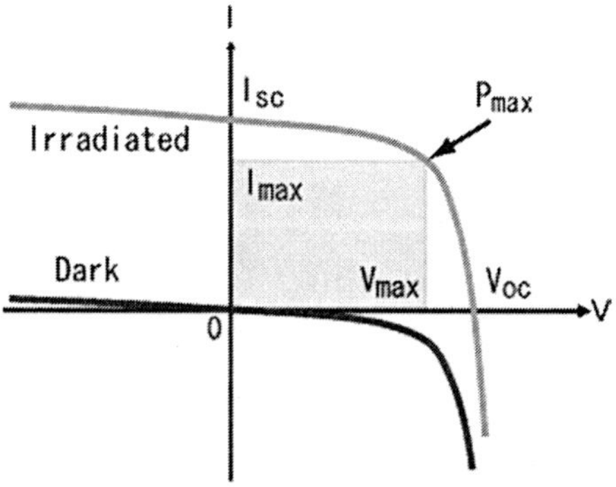

Figure : Schematic drawing of current-voltage characteristics of a solar cell The area of the yellow rectangle gives the output power. Pmax denotes the maximum power point

Residences and businesses that have a grid-tied electrical system are permitted in many countries to sell their energy to the utility grid. Electricity delivered to the grid can be compensated in several ways. "Net metering", is where the entity that owns the renewable energy power source receives compensation from the utility for its net outflow of power. So for example, if during a given month a power system feeds 500 kilowatt-hours into the grid and uses 100 kilowatt-hours from the grid, it would receive compensation for 400 kilowatt-hours. In the US, net metering policies vary by jurisdiction. Another policy is a feed-in tariff, where the producer is paid for every kilowatt hour delivered to the grid by a special tariff based on a contract with distribution company or other power authority. In the United States, grid-interactive power systems are covered by specific provisions in the National Electric Code, which also mandates certain requirements for grid-interactive inverters.

Typical Operation

Inverters take DC power and invert it to AC power so it can be fed into the electric utility company grid. The grid tie inverter must synchronise its frequency with that of the grid (e.g. 50 or 60 Hz) using a local oscillator and limit the voltage to no higher than the grid voltage. A high-quality modern GTI has a fixed unity power factor, which means its output voltage and current are perfectly lined up, and its phase angle is within 1 degree of the AC power grid. The inverter has an on-board computer which will sense the current AC grid waveform, and output a voltage to correspond with the grid.

Grid-tie inverters are also designed to quickly disconnect from the grid if the utility grid goes down. This is an NEC requirement that ensures that in the event of a blackout, the grid tie inverter will shut down to prevent the energy it produces from harming any line workers who are sent to fix the power grid. Properly configured, a grid tie inverter enables a home owner to use an alternative power generation system like solar or wind power without extensive rewiring and without batteries. If the alternative power being produced is insufficient, the deficit will be sourced from the electricity grid.

Technology

Grid-tie inverters that are available on the market today use a number of different technologies. The inverters may use the newer high-frequency transformers, conventional low-frequency transformers, or without transformer. Instead of converting direct current directly to 120 or 240 volts AC, high-frequency transformers employ a computerised multi-step process that involves converting the power to high-frequency AC and then back to DC and then to the final AC output voltage. Transformerless inverters, which boast lighter weight and higher efficiencies than their counterparts with transformers, are popular in Europe. However, transformerless inverters have been slow to enter the US market because historically there have concerns about having transformerless electrical systems feeding into the public utility grid. Since the lack of galvanic isolation between the DC and AC circuits could allow the passage of dangerous DC faults to be transmitted to the AC side. However, since 2005, the NFPA's NEC allows transformerless (or non-galvanically) inverters by removing the requirement that all solar electric systems to be negative grounded and specifying new safety requirements. The VDE 0126-1-1 and IEC 6210 also have been amended to allow and define the safety mechanisms needed for such systems. Primarily, residual or ground current detection is used to detect possible fault conditions. Also isolation tests are performed to insure DC to AC separation.

Most grid-tie inverters on the market include a maximum power point tracker on the input side that enables the inverter to extract a maximum amount of power from its intended power source. Since MPPT algorithms differ for solar panels and wind turbines, specially made inverters for each of these power sources are available.

Characteristics

Inverter manufacturers publish datasheets for the inverters in their product line. While the terminology and content will vary by manufacturer, datasheets generally include the information listed below.

- *Rated output power*: This value will be provided in watts or kilowatts. For some inverters, they may provide an output rating for different output voltages. For instance, if the inverter can be configured for either 240 VAC or 208 VAC output, the rated power output may be different for each of those configurations.
- *Output voltage(s)*: This value indicates to which utility voltages the inverter can connect. For smaller inverters that are designed for residential use, the output voltage is usually 240 VAC. Inverters that target commercial applications are rated for 208, 240, 277, 400, or 480 VAC and may also produce three phase power.
- *Peak efficiency*: The peak efficiency represents the highest efficiency that the inverter can achieve. Most grid-tie inverters on the market as of July 2009 have peak efficiencies of over 94%, some as high as 96%. The energy lost during inversion is for the most part converted into heat. This means that in order for an inverter to put out the rated amount of power it will need to have a power input that exceeds the output. For

example, a 5000 W inverter operating at full power at 95% efficiency will require an input of 5,263 W (rated power divided by efficiency). Inverters that are capable of producing power at different AC voltages may have different efficiencies associated with each voltage.

- *CEC weighted efficiency*: In contrast to peak efficiency, this value is an average efficiency and is a better representation of the inverter's operating profile. Inverters that are capable of producing power at different AC voltages may have different efficiencies associated with each voltage.
- *Maximum input current*: This is the maximum amount of direct current that the inverter will use. If a DC power source, such as a solar array, produces an amount of current that exceeds the maximum input current, that current will not be used by the inverter.
- *Maximum output current*: The maximum output current is the maximum continuous alternating current that the inverter will supply. This value is typically used to determine the minimum current rating of the over-current protection devices (e.g., breakers and fuses) and disconnects required for the output circuit. Inverters that are capable of producing power at different AC voltages will have different maximum outputs for each voltage.
- *Peak power tracking voltage*: This represents the DC voltage range in which the inverter's maximum point power tracker will operate. The system designer must configure the strings optimally so that during the majority of the year, the voltage of the strings will be within this range. This can be a difficult task since voltage will fluctuate with changes in temperature.
- *Start voltage*: This value is not listed on all inverter datasheets. The value indicates the minimum DC voltage that is required in order for the inverter to turn on and begin operation. This is especially important for solar applications, because the system designer must be sure that there is a sufficient number of solar modules wired in series in each string to produce this voltage. If this value is not provided by the manufacturer, system designers typically use the lower band of the peak power tracking voltage range as the inverter's minimum voltage.
- *IPxx rating*: The Ingress Protection rating or IP Code classifies and rates the level of protection provided against the ingress

of solid foreign objects (first digit) or water (second digit), a higher digit means greater protection. In the US the *NEMA rating* is used similarly to the international rating. Most inverters are rated for outdoors installation with IP45 (no dust protection) or IP65 (dust tight), or in the US, NEMA 3R (no windblown dust protection) or NEMA 4X (windblown dust, direct water splash and additional corrosion protection).

Quality of Service

A key concept of this system is the possibility of creating an electrical micro-system that is not dependent on the grid-tie to provide the high level of service quality which is often delivered by a traditional mains supply.

Conversely if the mains supply of the region is unreliable, the local generation system can be used to power important equipment.

Environmentally Friendly

Local generation can be from an environmentally friendly source such as photovoltaic. Individuals can choose to install their own system if an environmentally friendly mains provider is not available in their location.

B2G also spare the use of fossil fuel power plants and electrical substations to supply energy in peak loads of the public electric grid.

Small Scale Start

A micro generation facility can be started with a very small system such as a home wind power generation or photovoltaic (solar cells) generation system.

Sell to and Buy from Mains

- Excess electricity can be sold to mains.
- Electrical shortfall can be bought from mains.

List of Countries or Regions that Legally Allow Grid-tied Electrical Systems

- India
- Bangladesh
- Australia
- Belgium
- Bosnia and Herzegovina
- Canada

- Greece
- Hungary
- Israel
- Japan
- Mexico
- The Netherlands
- New Zealand
- Portugal
- Sri Lanka
- Singapore
- Spain
- USA
- Dominican Republic.

Stand-alone Power System

A stand-alone power system (SAPS or SPS), also known as remote area power supply (RAPS), is an off-the-grid electricity system for locations that are not fitted with an electricity distribution system. Typical SAPS include one or more methods of electricity generation, energy storage, and regulation.

Electricity is typically generated by one or more of the following methods:

- Solar panel
- Wind turbine
- Geothermal source.
- Micro combined heat and power.
- Micro hydro.
- Diesel or biofuel generator.

Storage is typically implemented as a battery bank, but other solutions exist including fuel cells. Power drawn directly from the battery will be direct current extra low voltage (DC ELV), and this is used especially for lighting as well as for DC appliances. An inverter is used to generate AC low voltage, which more typical appliances can be used with.

Stand-alone photovoltaic power systems are independent of the [[utility grid] and may use solar panels only or may be used in conjunction with a diesel generator or a wind turbine.

Ground (Electricity)

In electrical engineering, ground or earth is the reference point in an electrical circuit from which other voltages are measured, or is a common return path for electric current, or a direct physical connection to the Earth.

Figure : A typical earthing electrode (left) at a home in Australia.

Electrical circuits may be connected to ground (earth) for several reasons. In mains powered equipment, exposed metal parts are connected to ground to prevent contact with a dangerous voltage if electrical insulation fails. Connections to ground limit the build-up of static electricity when handling flammable products or when repairing electronic devices. In some telegraph and power transmission circuits, the earth itself can be used as one conductor of the circuit, saving the cost of installing a separate return conductor.

For measurement purposes, the Earth serves as a (reasonably) constant potential reference against which other potentials can be measured. An electrical ground system should have an appropriate current-carrying capability in order to serve as an adequate zero-voltage reference level.

In electronic circuit theory, a "ground" is usually idealised as an infinite source or sink for charge, which can absorb an unlimited amount of current without changing its potential. Where a real ground connection has a significant resistance, the approximation of zero potential is no longer valid. Stray voltages or earth potential rise effects will occur, which may create noise in signals or if large enough will produce an electric shock hazard.

The use of the term ground (or earth) is so common in electrical and electronics applications that circuits in portable electronic devices such as cell phones and media players as well as circuits in vehicles such as ships, aircraft, and spacecraft may be spoken of as having a "ground" connection without any actual connection to the Earth. This is usually a large conductor attached to one side of the power supply (such as the "ground plane" on a printed circuit board) which serves as the common return path for current from many different components in the circuit.

Synonyms

The terms *ground* and *grounding* are used in US electrical practice. In the UK the equivalent terms are *earth* and *earthing.*

The term is also mistakenly used for a bond wire or connection. A bond is any wire or attachment point that connects to the ground or earth. Meaning that what many refer to as a 'ground' wire or 'ground' screw is actually a 'bond' wire or screw.

History

Long-distance electromagnetic telegraph systems from 1820 onwards used two or more wires to carry the signal and return currents. It was then discovered, probably by the German scientist Carl August Steinheil in 1836-1837, that the ground could be used as the return path to complete the circuit, making the return wire unnecessary. However, there were problems with this system, exemplified by the transcontinental telegraph line constructed in 1861 by the Western Union Company between Saint Joseph, Missouri, and Sacramento, California. During dry weather, the ground connection often developed a high resistance, requiring water to be poured on the ground rod to enable the telegraph to work or phones to ring.

Later, when telephony began to replace telegraphy, it was found that the currents in the earth induced by power systems, electrical railways, other telephone and telegraph circuits, and natural sources including lightning caused unacceptable interference to the audio signals, and the two-wire system was reintroduced.

Radio Communications

An electrical connection to earth can be used as a reference potential for radio frequency signals for certain kinds of antennas. The part directly in contact with the earth-the "earth electrode"-can be as simple as a metal rod or stake driven into the earth, or a connection to buried metal water piping (though this carries the risk of the water pipe being later replaced with plastic). Because high frequency signals can flow to earth through capacitance, capacitance to ground is an important factor in effectiveness of signal grounds. Because of this a complex system of buried rods and wires can be effective. An ideal signal ground maintains zero voltage regardless of how much electric current flows into ground or out of ground.

The resistance at the signal frequency of the electrode-to-earth connection determines its quality, and that quality is improved by increasing the surface area of the electrode in contact with the earth, increasing the depth to which it is driven, using several connected ground rods, increasing the moisture of the soil, improving the conductive mineral content of the soil, and increasing the land area covered by the ground system.

Some types of transmitting antenna systems in the VLF, LF, MF and lower SW range depend on a good ground to operate efficiently. For example, a vertical monopole antenna requires a ground plane that often consists of an interconnected network of wires running radially away from the base of the antenna for a distance about equal to the height of the antenna. Sometimes such a ground plane is supported above ground to reduce losses.

AC Power Wiring Installations

In a mains electricity (AC power) wiring installation, the term ground conductor typically refers to three different conductors or conductor systems as listed below.

Equipment earthing conductors provide an electrical connection between non-current-carrying metallic parts of equipment and the earth. The reason for doing this according to the U.S. National Electrical Code (NEC), is to limit the voltage imposed by lightning, line surges, and contact with higher voltage lines. The equipment earthing conductor is usually also used as the equipment bonding conductor.

Equipment bonding conductors provide a low impedance path between non-current-carrying metallic parts of equipment and one of the conductors of that electrical system's source, so that if a part becomes energised for any reason, such as a frayed or damaged conductor, a short circuit will occur and operate a circuit breaker or fuse to disconnect the faulted circuit. Note that the earth itself has no role in this fault-clearing process since current must return to its source, not the earth as is sometimes believed.

By bonding (interconnecting) all exposed non-current carrying metal objects together, they should remain near the same potential thus reducing the chance of a shock. This is especially important in bathrooms where one may be in contact with several different metallic systems such as supply and drain pipes and appliance frames. The equipment bonding conductor is usually also used as the equipment earthing conductor.

A *grounding electrode conductor* connects one leg of an electrical system to one or more earth electrodes. This is called "system grounding" and most systems are required to be grounded. The U.S. NEC and the UK's BS 7671 list systems that are required to be grounded. The grounding electrode conductor is connected to the leg of the electrical system that is the "neutral wire". The grounding electrode conductor is also usually bonded to pipework and structural steel in larger structures. According to the NEC, the purpose of earthing an electrical system is to limit the voltage to earth imposed by lightning events and contact with higher voltage lines, and also to stabilise the voltage to earth during normal operation. In the past, water supply pipes were often used as ground electrodes, but this was banned where plastic pipes are popular. This type of ground applies to radio antennas and to lightning protection systems.

Permanently installed electrical equipment usually also has permanently connected grounding conductors. Portable electrical devices with metal cases may have them connected to earth ground by a pin in the interconnecting plug. The size of power ground conductors is usually regulated by local or national wiring regulations.

Power Transmission

In Single Wire Earth Return (SWER) AC electrical distribution systems, costs are saved by using just a single high voltage conductor for the power grid, while routing the AC return current through the earth. This system is mostly used in rural areas where large earth currents will not otherwise cause hazards.

Some HVDC power transmission systems use the ground as second conductor. This is especially common in schemes with submarine cables, as sea water is a good conductor. Buried grounding electrodes are used to make the connection to the earth. The site of these electrodes must be chosen very carefully in order to prevent electrochemical corrosion on underground structures.

A particular concern in design of electrical substations is earth potential rise. When very large fault currents are injected into the earth, the area around the point of injection may rise to a high potential with respect to distant points. This is due to the limited finite conductivity of the layers of soil in the earth. The gradient of the voltage (changing voltage within a distance) may be so high that two points on the ground may be at significantly different potentials, creating a hazard to anyone standing on the ground in the area. Pipes, rails, or communication wires entering a substation may see different ground potentials inside and outside the substation, creating a dangerous touch voltage.

Electronics

Signal grounds serve as return paths for signals and power (at extra low voltages, i.e., less than about 50 V) within equipment, and on the signal interconnections between equipment. Many electronic designs feature a single return that acts as a reference for all signals. Power and signal grounds often get connected together, usually through the metal case of the equipment.

Circuit Ground versus Earth

Voltage is a differential quantity. To measure the voltage of a single point, a reference point must be selected to measure against. This common reference point is called "ground" and considered to have zero voltage. This signal ground may not actually be connected to a power ground. A system where the system ground is not actually connected to another circuit or to earth (though there may still be AC coupling) is often referred to as a floating ground.

Separating Low Signal Ground from a Noisy Ground

In television stations, recording studios, and other installations where sound quality is critical, a special signal ground known as a "technical ground" (or "technical earth") is often installed, to prevent ground loops. This is basically the same thing as an AC power ground, but no appliance ground wires are allowed any connection to it, as they may carry electrical interference. In most cases, the studio's metal equipment racks are all joined together with heavy copper cables (or flattened copper tubing or busbars) and similar connections are made to the technical ground. Great care is taken that no AC-grounded appliances are placed on the racks, as a single AC ground connection to the technical ground will destroy its effectiveness. For particularly demanding applications, the main technical ground may consist of a heavy copper pipe, if necessary fitted by drilling through several concrete floors, such that all technical grounds may be connected by the shortest possible path to a grounding rod in the basement.

Lightning Protection Systems

Lightning protection systems are special grounding systems designed to safely conduct the extremely high voltage currents associated with lightning strikes.

Figure : Busbars are used for ground conductors in high-current circuits.

Earthing System

In electricity supply systems, an earthing (grounding) system defines the electrical potential of the conductors relative to that of the Earth's conductive surface. The choice of earthing system has implications for the safety and electromagnetic compatibility of the power supply. Note that regulations for earthing systems vary considerably between different countries. A functional earth connection serves a purpose other than providing protection against electrical shock. In contrast to a protective earth connection, a functional earth connection may carry a current during the normal operation of a device. Functional earth connections may be required by devices such as surge suppression and electromagnetic-compatibility filters, some types of antennas and various measurement instruments. Generally the protective earth is also used as a functional earth, though this requires care in some situations.

Ground (Earth) Mat

A ground (earth) mat or grounding (earthing) mat is a flat, flexible pad used for working on electrostatic sensitive devices. It is generally made of a conductive plastic or metal mesh covered substrate which is electrically attached to ground (earth). This helps discharge any static charge which a worker has built up, as well as any static charge on tools or exposed components laid on the mat. It is used most commonly in computer repair. Ground (earth) mats are also found on fuel trucks and tankers, which are otherwise insulated from ground (earth) as they make physical contact only with their tires and the air; obviously static discharge is undesirable during fuel-transfer operations. Similarly, in aircraft refuelling, a ground (earth) cable connects the tanker (truck or aeroplane) to the fuel-seeking craft to eliminate charge differences before fuel is transferred.

In an electrical substation a ground (earth) mat is a mesh of conductive material installed at places where a person would stand to operate a switch or other apparatus; it is bonded to the local supporting metal structure and to the handle of the switchgear, so that the operator will not be exposed to a high differential voltage due to a fault in the substation.

Harmonics (Electrical Power)

Harmonics are electric voltages and currents that appear on the electric power system as a result of certain kinds of electric loads. Harmonic frequencies in the power grid are a frequent cause of power quality problems.

Causes

In a normal alternating current power system, the voltage varies sinusoidally at a specific frequency, usually 50 or 60 hertz. When a linear electrical load is connected to the system, it draws a sinusoidal current at the same frequency as the voltage (though usually not in phase with the voltage). When a non-linear load, such as a rectifier, is connected to the system, it draws a current that is not necessarily sinusoidal. The current waveform can become quite complex, depending on the type of load and its interaction with other components of the system. Regardless of how complex the current waveform becomes, as described through Fourier series analysis, it is possible to decompose it into a series of simple sinusoids, which start at the power system fundamental frequency and occur at integer multiples of the fundamental frequency. Further examples of non-linear loads include common office equipment such as computers and printers, and also adjustable speed drives.

Effects

One of the major effects of power system harmonics is to increase the current in the system. This is particularly the case for the third harmonic, which causes a sharp increase in the zero sequence current, and therefore increases the current in the neutral conductor. This effect can require special consideration in the design of an electric system to serve non-linear loads. In addition to the increased line current, different pieces of electrical equipment can suffer effects from harmonics on the power system.

Motors

Electric motors experience hysteresis loss caused by eddy currents set up in the iron core of the motor. These are proportional to the frequency of the current. Since the harmonics are at higher frequencies, they produce more core loss in a motor than the power frequency would. This results in increased heating of the motor core, which (if excessive) can shorten the life of the motor. The 5th harmonic causes a CEMF (counter electromotive force) in large motors which acts in the opposite direction of rotation. The CEMF is not large enough to counteract the rotation, however it does play a small role in the resulting rotating speed of the motor.

Telephones

In the United States, common telephone lines are designed to transmit frequencies between 180 and 3200 Hz. Since electric power in the United States is distributed at 60 Hz, it normally does not interfere with telephone communications because its frequency is too low. However, since the third harmonic of the power has a frequency of 180 Hz, its higher-order harmonics are high enough to interfere with telephone service if they became induced in the line.

Phasor Measurement Unit

A phasor measurement unit (PMU) is a device which measures the electrical waves on an electricity grid, using a common time source for synchronisation. Time synchronisation allows synchronised real-time measurements of multiple remote measurement points on the grid. In power engineering, these are also commonly referred to as synchrophasors and are considered one of the most important measuring devices in the future of power systems. A PMU can be a dedicated device, or the PMU function can be incorporated into a protective relay or other device.

Technical Overview

A phasor is a complex number that represents both the magnitude and phase angle of the sine waves found in electricity. Phasor measurements that occur at the same time are called "synchrophasors", as are the PMU devices that allow their measurement. In typical applications phasor measurement units are sampled from widely dispersed locations in the power system network and synchronised from the common time source of a global positioning system (GPS) radio clock. Synchrophasor technology provides a tool for system operators and planners to measure the state of the electrical system and manage power quality. Synchrophasors measure voltages and currents at diverse locations on a power grid and can output accurately time-stamped voltage and current phasors. Because these phasors are truly synchronised, synchronised comparison of two quantities is possible, in real time. These comparisons can be used to assess system conditions.

The technology has the potential to change the economics of power delivery by allowing increased power flow over existing lines. Synchrophasor data could be used to allow power flow up to a line's dynamic limit instead of to its worst-case limit.

History

In 1893, Charles Proteus Steinmetz presented a paper on simplified mathematical description of the waveforms of alternating electricity. Steinmetz called his representation a phasor. With the invention of phasor measurement units (PMU) in 1988 by Dr. Arun G. Phadke and Dr. James S. Thorp at Virginia Tech, Steinmetz's technique of phasor calculation evolved into the calculation of real time phasor measurements that are synchronised to an absolute time reference provided by the Global Positioning System. Early prototypes of the PMU were built at Virginia Tech, and Macrodyne built the first PMU (model 1690) in 1992.

Phasor Networks

A phasor network consists of phasor measurement units (PMUs) dispersed throughout the electricity system, Phasor Data Concentrators (PDC) to collect the information and a Supervisory Control And Data Acquisition (SCADA) system at the central control facility. Such a network is used in Wide Area Measurement Systems (WAMS), the first of which was begun in 2000 by the Bonneville Power Administration. The complete network requires rapid data transfer within the frequency of sampling of the phasor data. GPS time stamping can provide a theoretical accuracy of synchronisation better than 1 microsecond. "Clocks need to be accurate to ± 500 nanoseconds to provide the one microsecond time standard needed by each device performing synchrophasor measurement." For 60 Hz systems, PMUs must deliver between 10 and 30 synchronous reports per second depending on the application. The PDC correlates the data, and controls and monitors the PMUs (from a dozen up to 60). At the central control facility, the SCADA system presents system wide data on all generators and substations in the system every 2 to 10 seconds. PMUs often use phone lines to connect to PDC, which then send data to the SCADA or Wide Area Measurement System (WAMS) server.

PMUs from multiple vendors can yield inaccurate readings. In one test, readings differed by 47 microseconds – or a difference of 1 degree of at 60 Hz- an unacceptable variance. China's solution to the problem was to build all its own PMUs adhering to its own specifications and standards so there would be no multi-vendor source of conflicts, standards, protocols, or performance characteristics.

Implementations

- The Bonneville Power Administration (BPA) is the first utility to implement comprehensive adoption of synchrophasors in its wide-area monitoring system. Today there are several implementations underway.
- The FNET project operated by Virginia Tech and the University of Tennessee utilises a network of approximately 80 low-cost, high-precision Frequency Disturbance Recorders to collect syncrophasor data from the U.S. power grid.
- In 2006, China's Wide Area Monitoring Systems (WAMS) for its 6 grids had 300 PMUs installed mainly at 500 kV and 330 kV substations and power plants. By 2012, China plans to have PMUs at all 500kV substations and all powerplants of 300MW and above. Since 2002, China has built its own PMUs to its own national standard. One type has higher sampling rates than typical and is used in power plants to measure rotor angle of the generator, reporting excitation voltage, excitation current, valve position, and output of the power system stabiliser (PSS). All PMUs are connected via private network, and samples are received within 40 ms on average.
- The North American Synchrophasor Initiative (NASPI), previously known as The Eastern Interconnect Phasor Project (EIPP), has over 120 connected phasor measurement units collecting data into a "Super Phasor Data Concentrator" system centred at Tennessee Valley Authority (TVA). This data concentration system is now an open source project known as the open PDC.
- The DOE has sponsored several related research projects, including GridStat at Washington State University.

Applications

1. Power system automation, as in smart grids
2. Load shedding and other load control techniques such as demand response mechanisms to manage a power system. (i.e. Directing power where it is needed in real-time)
3. Increase the reliability of the power grid by detecting faults early, allowing for isolation of operative system, and the prevention of power outages.
4. Increase power quality by precise analysis and automated correction of sources of system degradation.
5. Wide Area measurement and control, in very wide area super grids, regional transmission networks, and local distribution grids.

Standards

The IEEE 1344 standard for synchrophasors was completed in 1995, and reaffirmed in 2001. In 2005, it was replaced by IEEE Standard C37.118-2005, which was a complete revision and dealt with issues concerning use of PMUs in electric power systems. The specification describes standards for measurement, the method of quantifying the measurements, testing & certification requirements for verifying accuracy, and data transmission format and protocol for real-time data communication. The standard is not yet comprehensive-it does not attempt to address all factors that PMUs can detect in power system dynamic activity.

Other standards used with PMU interfacing:

- OPC-DA / OPC-HDA - A Microsoft Windows based interface protocol that is currently being generalised to use XML and run on non Windows computers.
- IEC 61850 a standard for electrical substation automation

BPA PDCStream - a variant of IEEE 1344 used by the Bonneville Power Administration (BPA) PDCs and user interface software.

Parasitic Load

Parasitic load is a term used with regard to electrical appliances and railway locomotives.

With regard to electrical appliances, it represents the power consumed even when the appliance is shut off, that is standby power.

With regard to railway locomotives, it is any of the loads or devices powered by the diesel prime mover not contributing to tractive effort, such as air compressor, traction motor blower, or radiator fans.

With regard to electricity production, it is any of the loads or devices powered by the wind generator, not contributing to net electric yield found by subtracting productive yield from gross yield or

GY - PY = PL where

"GY" is gross electric yield (the output of the generator); "PY" is productive yield (the electricity which is made available to external electric loads);and "PL" is parasitic load.

Chapter 7

Polyphase System

A polyphase system is a means of distributing alternating current electrical power. Polyphase systems have three or more energised electrical conductors carrying alternating currents with a definite time offset between the voltage waves in each conductor. Polyphase systems are particularly useful for transmitting power to electric motors. The most common example is the three-phase power system used for industrial applications and for power transmission. The most obvious advantage of three-phase power transmission using three wires, as compared to single-phase power transmission over two wires, is that the power transmitted in the three-phase system is the voltage multiplied by the current in each wire times the square root of three (approximately 1.73). The power transmitted by the single-phase system is simply the voltage multiplied by the current. Thus the three-phase system transmits 73% more power but uses only 50% more wire.

Phases

In the very early days of commercial electric power, some installations used two-phase four-wire systems for motors. The chief advantage of these was that the winding configuration was the same as for a single-phase capacitor-start motor and, by using a four-wire system, conceptually the phases were independent and easy to analyse with mathematical tools available at the time. Two-phase systems can also be implemented using three wires (two "hot" plus a common neutral). However this introduces asymmetry; the voltage drop in the neutral makes the phases not exactly 90 degrees apart. Two-phase systems have been replaced with three-phase systems. A two-phase supply with 90 degrees between phases can be derived from a three-phase system using a Scott-connected transformer.

A polyphase system must provide a defined direction of phase rotation, so mirror image voltages do not count towards the phase order. A 3-wire system with two phase conductors 180 degrees apart is still only single phase. Such systems are sometimes described as split-phase.

Motors

Polyphase power is particularly useful in AC motors, such as the induction motor, where it generates a rotating magnetic field. When a three-phase supply completes one full cycle, the magnetic field of a two-pole motor has rotated through 360° in physical space; motors with more pairs of poles require more power supply cycles to complete one physical revolution of the magnetic field and so these motors run slower. Nikola Tesla and Michail Dolivo-Dobrovolsky invented the first practical induction motors using a rotating magnetic field-previously all commercial motors were DC, with expensive commutators, high-maintenance brushes and characteristics unsuitable for operation on an alternating current network. Polyphase motors are simple to construct, are self-starting and have little vibration compared with single-phase motors.

Higher Phase Order

Higher phase numbers than three have been used. A common practice for rectifier installations and in HVDC converters is to provide six phases, with 60 degree phase spacing, to reduce harmonic generation in the AC supply system and to provide smoother direct current. Experimental high-phase-order transmission lines have been built with up to 12 phases. These allow application of Extra High Voltage (EHV) design rules at lower voltages and would permit increased power transfer in the same transmission line corridor width.

Fuzzy Set

Fuzzy sets are sets whose elements have degrees of membership. Fuzzy sets were introduced simultaneously by Lotfi A. Zadeh and Dieter Klaua in 1965 as an extension of the classical notion of set. In classical set theory, the membership of elements in a set is assessed in binary terms according to a bivalent condition — an element either belongs or does not belong to the set. By contrast, fuzzy set theory permits the gradual assessment of the membership of elements in a set; this is described with the aid of a membership function valued in the real unit interval [0, 1]. Fuzzy sets generalise classical sets, since the indicator functions of classical sets are special cases of the membership functions of fuzzy sets, if the latter only take values 0 or 1. In fuzzy set theory, classical bivalent sets are usually called *crisp* sets. The fuzzy set theory can be used in a wide range of domains in which information is incomplete or imprecise, such as bioinformatics. Fuzzy sets can be applied, for example, to the field of genealogical research. When an individual is searching in vital records such as birth records for possible ancestors, the researcher must contend with a number of issues that could be encapsulated in a membership function. Looking for an ancestor named John Henry Pittman, who you *think* was born in (probably eastern) Tennessee circa 1853 (based on statements of his age in later censuses, and a marriage record in Knoxville), what is the likelihood that a particular birth record for "John Pittman" is *your* John Pittman? What about a record in a different part of Tennessee for "J.H. Pittman" in 1851? (It has been suggested by Thayer Watkins that Zadeh's ethnicity is an example of a fuzzy set)

Definition

A fuzzy set is a pair *(U, m)* where *U* is a set and $m: U \rightarrow [0,1]$.

For each $x \in U$, the value *m(x)* is called the grade of membership of *x* in *(U, m)* For a finite set $U = \{x_1, \ldots, x_n\}$, the fuzzy set *(U, m)* is often denoted by

Let $x \in U$. Then *x* is called not included in the fuzzy set *(U, m)* if *m(x) = 0* *x* is called fully included if *m(x) = 1* and *x* is called a fuzzy member if $0 < m(x) < 1$. The set $\{x \in U \mid m(x) > 0\}$ is called the support of *(U, m)* and the set $\{x \in U \mid m(x) = 1\}$ is called its kernel. The function $m$is called the membership function of the fuzzy set *(U, m)*.

Sometimes, more general variants of the notion of fuzzy set are used, with membership functions taking values in a (fixed or variable) algebra or structure *L* of a given kind; usually it is required that *L* be at least a poset or lattice. These are usually called *L*-fuzzy sets, to distinguish them from those valued over the unit interval. The usual membership functions with values in [0, 1] are then called [0, 1]-valued membership functions. These kinds of generalisations were first considered in 1967 by Joseph Goguen, who was a student of Zadeh.

Fuzzy Logic

Fuzzy logic is a form of many-valued logic; it deals with reasoning that is approximate rather than fixed and exact. In contrast with traditional logic theory, where binary sets have two-valued logic: true or false, fuzzy logic variables may have a truth value that ranges in degree between 0 and 1. Fuzzy logic has been extended to handle the concept of partial truth, where the truth value may range between completely true and completely false. Furthermore, when linguistic variables are used, these degrees may be managed by specific functions.

Fuzzy logic began with the 1965 proposal of fuzzy set theory by Lotfi Zadeh. Fuzzy logic has been applied to many fields, from control theory to artificial intelligence.

Overview

The reasoning in fuzzy logic is similar to human reasoning. It allows for approximate values and inferences as well as incomplete or ambiguous data (fuzzy data) as opposed to only relying on crisp data (binary yes/no choices). Fuzzy logic is able to process incomplete data and provide approximate solutions to problems other methods find difficult to solve. Terminology used in fuzzy logic not used in other methods are: very high, increasing, somewhat decreased, reasonable and very low.

Degrees of Truth

Fuzzy logic and probabilistic logic are mathematically similar – both have truth values ranging between 0 and 1 – but conceptually distinct, due to different interpretations. Fuzzy logic corresponds to "degrees of truth", while probabilistic logic corresponds to "probability, likelihood"; as these differ, fuzzy logic and probabilistic logic yield different models of the same real-world situations. Both degrees of truth and probabilities range between 0 and 1 and hence may seem similar at first. For example, let a 100 ml glass contain 30 ml of water. Then we may consider two concepts: Empty and Full. The meaning of each of them can be represented by a certain fuzzy set. Then one might define the glass as being 0.7 empty and 0.3 full. Note that the concept of emptiness would be subjective and thus would depend on the observer or designer. Another designer might equally well design a set membership function where the glass would be considered full for all values down to 50 ml. It is essential to realise that fuzzy logic uses truth degrees as a mathematical model of the vagueness phenomenon while probability is a mathematical model of ignorance.

Applying Truth Values

A basic application might characterise subranges of a continuous variable. For instance, a temperature measurement for anti-lock brakes might have several separate membership functions defining particular temperature ranges needed to control the brakes properly. Each function maps the same temperature value to a truth value in the 0 to 1 range. These truth values can then be used to determine how the brakes should be controlled.

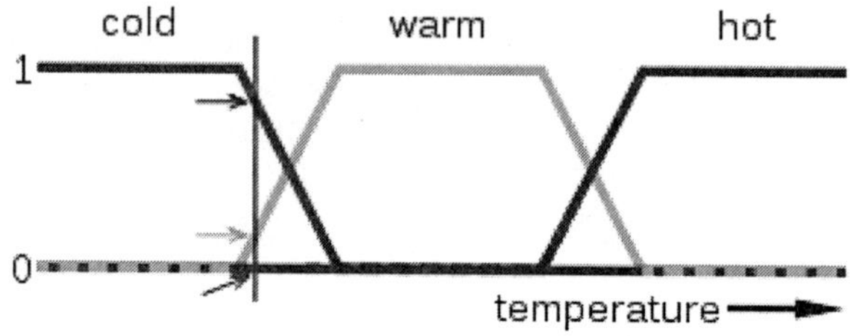

Figure : Fuzzy logic temperature

The meanings of the expressions *cold*, *warm*, and *hot* are represented by functions mapping a temperature scale. A point on that scale has three "truth values"—one for each of the three functions. The vertical line in the image represents a particular temperature that the three arrows (truth values) gauge. Since the red arrow points to zero, this temperature may be interpreted as "not hot". The orange arrow (pointing at 0.2) may describe it as "slightly warm" and the blue arrow (pointing at 0.8) "fairly cold".

Linguistic Variables

While variables in mathematics usually take numerical values, in fuzzy logic applications, the non-numeric *linguistic variables* are often used to facilitate the expression of rules and facts.

A linguistic variable such as *age* may have a value such as *young* or its antonym *old*. However, the great utility of linguistic variables is that they can be modified via linguistic hedges applied to primary terms. The linguistic hedges can be associated with certain functions.

Example

Fuzzy set theory defines fuzzy operators on fuzzy sets. The problem in applying this is that the appropriate fuzzy operator may not be known. For this reason, fuzzy logic usually uses IF-THEN rules, or constructs that are equivalent, such as fuzzy associative matrices.

Rules are usually expressed in the form:

IF *variable* IS *property* THEN *action*

For example, a simple temperature regulator that uses a fan might look like this:

IF temperature IS very cold THEN stop fan

IF temperature IS cold THEN turn down fan

IF temperature IS normal THEN maintain level

IF temperature IS hot THEN speed up fan.

There is no "ELSE" – all of the rules are evaluated, because the temperature might be "cold" and "normal" at the same time to different degrees.

The AND, OR, and NOT operators of boolean logic exist in fuzzy logic, usually defined as the minimum, maximum, and complement; when they are defined this way, they are called the *Zadeh operators*. So for the fuzzy variables x and y:

NOT x = (1 - truth(x))

x AND y = minimum(truth(x), truth(y))

x OR y = maximum(truth(x), truth(y))

There are also other operators, more linguistic in nature, called *hedges* that can be applied. These are generally adverbs such as "very", or "somewhat", which modify the meaning of a set using a mathematical formula.

Logical Analysis

In mathematical logic, there are several formal systems of "fuzzy logic"; most of them belong among so-called t-norm fuzzy logics.

Propositional Fuzzy Logics

The most important propositional fuzzy logics are:

- Monoidal t-norm-based propositional fuzzy logic MTL is an axiomatisation of logic where conjunction is defined by a left continuous t-norm, and implication is defined as the residuum of the t-norm. Its models correspond to MTL-algebras that are prelinear commutative bounded integral residuated lattices.
- Basic propositional fuzzy logic BL is an extension of MTL logic where conjunction is defined by a continuous t-norm, and implication is also defined as the residuum of the t-norm. Its models correspond to BL-algebras.
- £ukasiewicz fuzzy logic is the extension of basic fuzzy logic BL where standard conjunction is the £ukasiewicz t-norm. It has the axioms of basic fuzzy logic plus an axiom of double negation, and its models correspond to MV-algebras.
- Gödel fuzzy logic is the extension of basic fuzzy logic BL where conjunction is Gödel t-norm. It has the axioms of BL plus an axiom of idempotence of conjunction, and its models are called G-algebras.

- Product fuzzy logic is the extension of basic fuzzy logic BL where conjunction is product t-norm. It has the axioms of BL plus another axiom for cancellativity of conjunction, and its models are called product algebras.
- Fuzzy logic with evaluated syntax (sometimes also called Pavelka's logic), denoted by EV£, is a further generalisation of mathematical fuzzy logic. While the above kinds of fuzzy logic have traditional syntax and many-valued semantics, in EV£ is evaluated also syntax. This means that each formula has an evaluation. Axiomatisation of EV£ stems from £ukasziewicz fuzzy logic. A generalisation of classical Gödel completeness theorem is provable in EV£.

Predicate Fuzzy Logics

These extend the above-mentioned fuzzy logics by adding universal and existential quantifiers in a manner similar to the way that predicate logic is created from propositional logic. The semantics of the universal (resp. existential) quantifier in t-norm fuzzy logics is the infimum (resp. supremum) of the truth degrees of the instances of the quantified subformula.

Decidability Issues for Fuzzy Logic

The notions of a "decidable subset" and "recursively enumerable subset" are basic ones for classical mathematics and classical logic. Then, the question of a suitable extension of such concepts to fuzzy set theory arises. A first proposal in such a direction was made by E.S. Santos by the notions of *fuzzy Turing machine, Markov normal fuzzy algorithm* and *fuzzy program*. Successively, L. Biacino and G. Gerla showed that such a definition is not adequate and therefore proposed the following one. $\ddot{U}$ denotes the set of rational numbers in [0,1]. A fuzzy subset $s : S \rightarrow [0,1]$ of a set S is *recursively enumerable* if a recursive map $h : S{\times}N \rightarrow \ddot{U}$ exists such that, for every x in S, the function $h(x,n)$ is increasing with respect to n and $s(x) = \lim h(x,n)$. We say that s is *decidable* if both s and its complement $-s$ are recursively enumerable. An extension of such a theory to the general case of the L-subsets is proposed in Gerla 2006. The proposed definitions are well related with fuzzy logic. Indeed, the following theorem holds true (provided that the deduction apparatus of the fuzzy logic satisfies some obvious effectiveness property).

Theorem. Any axiomatizable fuzzy theory is recursively enumerable. In particular, the fuzzy set of logically true formulas is recursively enumerable in spite of the fact that the crisp set of valid formulas is not recursively enumerable, in general. Moreover, any axiomatizable and complete theory is decidable.

It is an open question to give supports for a *Church thesis* for fuzzy logic claiming that the proposed notion of recursive enumerability for fuzzy subsets is the adequate one. To this aim, further investigations on the notions of fuzzy grammar and fuzzy Turing machine should be necessary. Another open question is to start from this notion to find an extension of Gödel's theorems to fuzzy logic.

Fuzzy Databases

Once fuzzy relations are defined, it is possible to develop fuzzy relational databases. The first fuzzy relational database, FRDB, appeared in Maria Zemankova's dissertation. Later, some other models arose like the Buckles-Petry model, the Prade-Testemale Model, the Umano-Fukami model or the GEFRED model by J.M. Medina, M.A. Vila et al. In the context of fuzzy databases, some fuzzy querying languages have been defined, highlighting the SQLf by P. Bosc et al. and the FSQL by J. Galindo et al. These languages define some structures in order to include fuzzy aspects in the SQL statements, like fuzzy conditions, fuzzy comparators, fuzzy constants, fuzzy constraints, fuzzy thresholds, linguistic labels and so on.

Comparison to Probability

Fuzzy logic and probability are different ways of expressing uncertainty. While both fuzzy logic and probability theory can be used to represent subjective belief, fuzzy set theory uses the concept of fuzzy set membership (i.e., *how much* a variable is in a set), and probability theory uses the concept of subjective probability (i.e., *how probable* do I think that a variable is in a set). While this distinction is mostly philosophical, the fuzzy-logic-derived possibility measure is inherently different from the probability measure, hence they are not *directly* equivalent. However, many statisticians are persuaded by the work of Bruno de Finetti that only one kind of mathematical uncertainty is needed and thus fuzzy logic is unnecessary. On the other hand, Bart Kosko argues that probability is a subtheory of fuzzy logic, as probability only handles one kind of uncertainty. He also claims to have proven a derivation of Bayes' theorem from the concept of fuzzy subsethood. Lotfi Zadeh argues that fuzzy logic is different in character from probability, and is not a replacement for it. He fuzzified probability to fuzzy probability and also generalised it to what is called possibility theory.

Fuzzy Number

A fuzzy number is an extension of a regular number in the sense that it does not refer to one single value but rather to a connected set of possible values, where each possible value has its own weight between 0 and 1. This weight is called the membership function. A fuzzy number is thus a special case of a convex fuzzy set. Just like Fuzzy logic is an extension of Boolean logic (which uses absolute truth and falsehood only, and nothing in between), fuzzy numbers are an extension of real numbers. Calculations with fuzzy numbers allow the incorporation of uncertainty on parameters, properties, geometry, initial conditions, etc.

Fuzzy Interval

A fuzzy interval is an uncertain set $\widetilde{A} \subseteq \mathbb{R}$ with a mean interval whose elements possess the membership function value $\mu_A(x) = 1$. As in fuzzy numbers, the membership function must be convex, normalised, at least segmentally continuous.

Fuzzy Relation Equation

The fuzzy relation equation is an equation of the form A · R = B, where A and B are fuzzy sets, R is a fuzzy relation, and A · R stands for the composition of A with R.

Electricity Liberalisation

Electricity liberalisation refers to the liberalisation of electricity markets. As electricity supply is a natural monopoly, this entails complex and costly systems of regulation to enforce a system of competition.

The standard model for electricity liberalisation is the "British model", after the British system which began in the late 1980s with the privatization and vertical de-integration of the United Kingdom electricity industry. At a cost of over $3bn, a system of competition was developed to auction spare capacity through a central system. The risks involved for both generators and distributors have led to vertical re-integration.

Liberalisation of electricity tends to substantially benefit large consumers (mainly industrial users), but benefits for domestic consumers compared with a public monopoly or a regulated private monopoly are questionable. There are also doubts over whether the system can ensure long-term security of supply through providing sufficient incentives to begin building generation capacity in time for when it is needed. Electricity markets in the European Union are governed by the Internal Market in Electricity directive.

Demand Response

In electricity grids, demand response (DR) is similar to dynamic demand mechanisms to manage customer consumption of electricity in response to supply conditions, for example, having electricity customers reduce their consumption at critical times or in response to market prices. The difference is that demand response mechanisms respond to explicit requests to shut off, whereas dynamic demand devices passively shut off when stress in the grid is sensed. Demand response can involve actually curtailing power used or by starting on site generation which may or may not be connected in parallel with the grid. This is a quite different concept from energy efficiency, which means using less power to perform the same tasks, on a continuous basis or whenever that task is performed. At the same time, demand response is a component of smart energy demand, which also includes energy efficiency, home and building energy management, distributed renewable resources, and electric vehicle charging.

Current demand response schemes are implemented with large and small commercial as well as residential customers, often through the use of dedicated control systems to shed loads in response to a request by a utility or market price conditions. Services (lights, machines, air conditioning) are reduced according to a preplanned load prioritisation scheme during the critical time frames. An alternative to load shedding is on-site generation of electricity to supplement the power grid. Under conditions of tight electricity supply, demand response can significantly decrease the peak price and, in general, electricity price volatility.

Demand response is generally used to refer to mechanisms used to encourage consumers to reduce demand, thereby reducing the peak demand for electricity. Since electrical generation and transmission systems are generally sized to correspond to peak demand (plus margin for forecasting error and unforeseen events), lowering peak demand reduces overall plant and capital cost requirements. Depending on the configuration of generation capacity, however, demand response may also be used to increase demand (load) at times of high production and low demand. Some systems may thereby encourage energy storage to arbitrage between periods of low and high demand (or low and high prices).

There are three types of demand response-emergency demand response, economic demand response and ancillary services demand response. Emergency demand response is employed to avoid involuntary service interruptions during times of supply scarcity. Economic demand response is employed to allow electricity customers to curtail their consumption when the productive or convenience of consuming that electricity is worth less to them than paying for the electricity.

Ancillary services demand response consists of a number of speciality services that are needed to ensure the secure operation of the transmission grid and which have traditionally been provided by generators.

Smart Grid Application

Smart grid applications improve the ability of electricity producers and consumers to communicate with one another and make decisions about how and when to produce and consume kWh.

This emerging technology will allow customers to shift from an event based demand response where the utility requests the shedding of load, towards a more 24/7 based demand response where the customer sees incentives for controlling load all the time. Although this back and forth dialogue increases the opportunities for demand response, customers are still largely influenced by economic incentives and are reluctant to relinquish total control of their assets to utility companies.

One advantage of a smart grid application is time-based pricing. Customers who traditionally pay a fixed rate for kWh and kW/month can set their threshold and adjust their usage to take advantage of fluctuating prices.

This may require the use of an energy management system to control appliances and equipment and can involve economies of scale. Another advantage, mainly for large customers with generation, is being able to closely monitor, shift, and balance load in a way that allows the customer to save peak load and not only save on kWh and kW/month but be able to trade what they have saved in an energy market. Again this involves sophisticated energy management systems, incentives, and a viable trading market.

Smart grid applications increase the opportunities for demand response by providing real time data to producers and consumers, but the economic and environmental incentives remain the driving force behind the practice.

Electricity Pricing

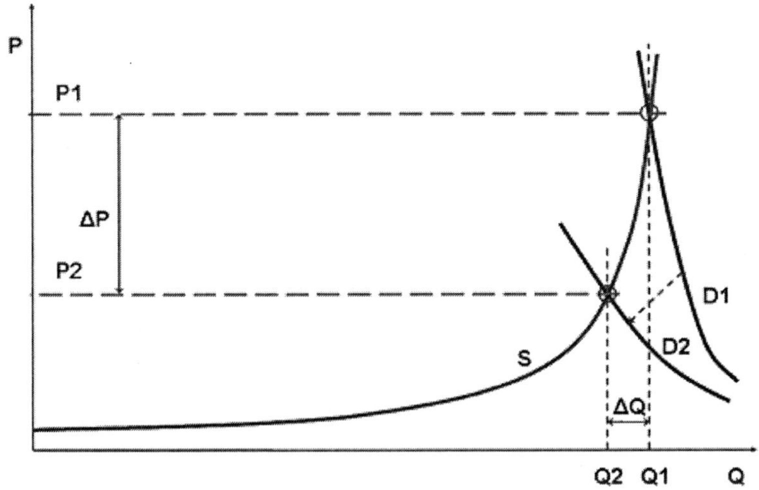

Figure : Explanation of demand response effects on a quantity (Q) - price (P) graph. Under inelastic demand (D1) extremely high price (P1) may result on a strained electricity market.

If demand response measures are employed the demand becomes more elastic (D2). A much lower price will result in the market (P2).

It is estimated that a 5% lowering of demand would result in a 50% price reduction during the peak hours of the California electricity crisis in 2000/2001. The market also becomes more resilient to intentional withdrawal of offers from the supply side.

In most electric power systems, some or all consumers pay a fixed price per unit of electricity independent of the cost of production at the time of consumption. The consumer price may be established by the government or a regulator, and typically represents an average cost per unit of production over a given timeframe (for example, a year). Consumption therefore is not sensitive to the cost of production in the short term (e.g. on an hourly basis). In economic terms, consumers' usage of electricity is inelastic in short time frames since the consumers do not face the actual price of production; if consumers were to face the short run costs of production they would generally increase and decrease their use of electricity in reaction to those cost-based price signals. In effect, consumers served under these fixed rate tariffs are endowed with real "call options" on electricity.

In virtually all power systems electricity is produced by generators that are dispatched in merit order, i.e., generators with the lowest marginal cost (lowest variable cost of production) are used first, followed by the next cheapest, etc., until the instantaneous electricity demand is satisfied. In most power systems the wholesale price of electricity will be equal to the marginal cost of the highest cost generator that is injecting energy, which will vary with the level of demand. Thus the variation in pricing can be significant: for example, in Ontario between August and September 2006, wholesale prices (in Canadian Dollars) paid to producers ranged from a peak of $318 per MW·h to a minimum of-(negative) $3.10 per MW·h.

It is not unusual for the price to vary by a factor of two to five due to the daily demand cycle. A negative price indicates that producers were being charged to provide electricity to the grid (and consumers paying real-time pricing may have actually received a rebate for consuming electricity during this period). This generally occurs at night when demand falls to a level where all generators are operating at their minimum output levels and some of them must be shut down. The negative price is the inducement to bring about these shutdowns in a least-cost manner.

Two Carnegie Mellon studies in 2006 looked at the importance of demand response for the electricity industry in general terms and with specific application of real-time pricing for consumers for the PJM Interconnection Regional Transmission authority. The latter study found that even small shifts in peak demand would have a large effect on savings to consumers and avoided costs for additional peak capacity: a 1% shift in peak demand would result in savings of 3.9%, billions of dollars at the system level. An approximately 10% reduction in peak demand (achievable depending on the elasticity of demand) would result in systems savings of between $8 to $28 billion.

In a discussion paper, Ahmad Faruqui, a principal with the Brattle Group, estimates that a 5 percent reduction in US peak electricity demand could produce approximately $35 billion in cost savings over a 20 year period, exclusive of the cost of the metering and communications needed to implement the dynamic pricing needed to achieve these reductions. While the net benefits would be significantly less than the claimed $35 billion, they would still be quite substantial. In Ontario, Canada, the Independent Electricity System Operator has noted that in 2006, peak demand exceeded 25,000 megawatts during only 32 system hours (less than 0.4% of the time), while maximum demand during the year was just over 27,000 megawatts. The ability to "shave" peak demand based on reliable commitments would therefore allow the province to reduce built capacity by approximately 2,000 megawatts.

Electricity Grids and Peak Demand Response

In an electricity grid, electricity consumption and production must balance at all times; any significant imbalance could cause grid instability or severe voltage fluctuations, and cause failures within the grid. Total generation capacity is therefore sized to correspond to total peak demand with some margin of error and allowance for contingencies (such as plants being off-line during peak demand periods). Operators will generally plan to use the least expensive generating capacity (in terms of marginal cost) at any given period, and use additional capacity from more expensive plants as demand increases. Demand response in most cases is targeted at reducing peak demand to reduce the risk of potential disturbances, avoid additional capital cost requirements for additional plant, and avoid use of more expensive and/or less efficient operating plant. Consumers of electricity will also pay lower prices if generation capacity that would have been used is from a higher-cost source of power generation.

Demand response may also be used to increase demand during periods of high supply and/or low demand. Some types of generating plant must be run at close to full capacity (such as nuclear), while other types may produce at negligible marginal cost (such as wind and solar). Since there is usually limited capacity to store energy, demand response may attempt to increase load during these periods to maintain grid stability. For example, in the province of Ontario in September 2006, there was a short period of time when electricity prices were negative for certain users. Energy storage such as Pumped-storage hydroelectricity is a way to increase load during periods of low demand for use during later periods. Use of demand response to increase load is less common, but may be necessary or efficient in systems where there are large amounts of generating capacity that cannot be easily cycled down.

Some grids may use pricing mechanisms that are not real-time, but easier to implement (users pay higher prices during the day and lower prices at night, for example) to provide some of the benefits of the demand response mechanism with less demanding technological requirements. For example, in 2006 Ontario began implementing a "Smart Meter" program that implements "Time-of-Use" (TOU) pricing, which tiers pricing according to on-peak, mid-peak and off-peak schedules. During the winter, on-peak is defined as morning and early evening, mid-peak as mid-day to late afternoon, and off-peak as night-time; during the summer, the on-peak and mid-peak periods are reversed, reflecting air conditioning as the driver of summer demand.

In 2007, prices during the off-peak were C$0.034 per KWh and C$0.097 during the on-peak demand period, or just less than three times as expensive. As of 2007, few utilities had the meters and systems capability to implement TOU pricing, however, and most customers are not expected to get smart meters until 2008-2010. Eventually, the TOU pricing (or real-time pricing) is expected to be mandatory for most customers in the province.

Load Shedding

Electrical generation and transmission systems may not always meet peak demand requirements— the greatest amount of electricity required by all utility customers within a given region. In the effort to reduce the electric demand on power grids at critical periods, researchers developed a ballast prototype that quickly and reliably sheds the electric load within a building's lighting system. A load-shedding ballast is an instant-start ballast with bi-level dimming and a built-in power line carrier (PLC) signal receiver for automated dimming response.

By dimming lighting via an electronic signal, the ballast reduces the current supplied to the lamps. A signal injector on the building's lighting circuits controls the ballasts, eliminating the need for extra wiring. The ballasts respond to a signal sent by the utility or the customer's energy management system, reducing power to the lighting by one third. Field studies showed that building owners could dim the lights by as much as 40% for brief periods of time without upsetting 70% of the building's occupants or hindering productivity. Ninety percent of building occupants accepted the reduction in light levels when they were told that it was being done to conserve energy.

The new ballast system works on individual light fixtures, not on the main power grid. The system is recommended for new construction and remodelling and promises good return on investment in energy savings. In U.S.-based markets, the system has a three-year or less payback period for new construction. The ballast's use has the potential to reduce U.S. peak electric demand by 20,000 megawatts. If used widely, it has the potential to help avoid blackouts.

Incentives to Shed Loads

Energy consumers need some incentive to respond to such a request from a Demand Response Provider. Demand Response incentives can be formal or informal. For example, the utility might create a tariff-based incentive by passing along short-term increases in the price of electricity. Or they might impose mandatory cutbacks during a heat wave for selected high-volume users, who are compensated for their participation. Other users may receive a rebate or other incentive based on firm commitments to reduce power during periods of high demand, sometimes referred to as *negawatts*.

Commercial and industrial power users might impose load shedding on themselves, without a request from the utility. Some businesses generate their own power and wish to stay within their energy production capacity to avoid buying power from the grid. Some utilities have commercial tariff structures that set a customer's power costs for the month based on the customer's moment of highest use, or peak demand. This encourages users to flatten their demand for energy, known as energy demand management, which sometimes requires cutting back services temporarily.

Smart metering has been implemented in some jurisdictions to provide real-time pricing for all types of users, as opposed to fixed-rate pricing throughout the demand period. In this application, users have a direct incentive to reduce their use at high-demand, high-price periods. Many users may not be able to effectively reduce their demand at various times, or the peak prices may be lower than the level required to induce a change in demand during short time periods (users have low price sensitivity, or elasticity of demand is low). Automated control systems exist, which, although effective, may be too expensive to be feasible for some applications.

Technologies for Demand Reduction

Technologies are available, and more are under development, to automate the process of demand response. Such technologies detect the need for load shedding, communicate the demand to participating users, automate load shedding, and verify compliance with demand-response programs. GridWise and EnergyWeb are two major federal initiatives in the United States to develop these technologies.

Universities and private industry (including EnergyConnect, Inc., Energy Curtailment Specialists, North America Power Partners, Comverge, Inc., EnerNOC, Inc., GridPoint, Inc., CPower, Inc., Site-Controls, LLC., Powerit Solutions, RTP Controls, Inc., Fifth Fuel Management, and Energy Optimisers Ltd (Plogg)) are also doing research and development in this arena. Scalable and comprehensive software solutions for DR (such as platforms by Ziphany, LLC and Convia, Inc./A Herman Miller Company, and Servidyne via its iTendant offering) enable business and industry growth.

Some utilities are considering and testing automated systems connected to industrial, commercial and residential users that can reduce consumption at times of peak demand, essentially delaying draw marginally. Although the amount of demand delayed may be small, the implications for the grid (including financial) may be substantial, since system stability planning often involves building capacity for extreme peak demand events, plus a margin of safety in reserve. Such events may only occur a few times per year.

The process may involve turning down or off certain appliances or sinks (and, when demand is unexpectedly low, potentially increasing usage). For example, heating may be turned down or air conditioning or refrigeration may be turned up (turning up to a higher temperature uses less electricity), delaying slightly the draw until a peak in usage has passed. In the city of Toronto, certain residential users can participate in a program (Peaksaver AC) whereby the system operator can automatically control hot water heaters or air conditioning during peak demand; the grid benefits by delaying peak demand (allowing peaking plants time to cycle up or avoiding peak events), and the participant benefits by delaying consumption until after peak demand periods, when pricing should be lower. Although this is an experimental program, at scale these solutions have the potential to reduce peak demand considerably. The success of such programs depends on the development of appropriate technology, a suitable pricing system for electricity, and the cost of the underlying technology. Bonneville Power experimented with direct-control technologies in Washington and Oregon residences, and found that the avoided transmission investment would justify the cost of the technology.

Other methods to implementing demand response approach the issue of subtlely reducing duty cycles rather than implementing thermostat setbacks. These can be implemented using customised building automation systems programming, or through swarm-logic methods coordinating multiple loads in a facility (e.g. REGEN Energy's EnviroGrid controllers).

It was recently announced that electric refrigerators will be sold in the UK fitted with a frequency sensing device which will delay or advance the cooling cycle based on monitoring grid frequency.

Short-term Inconvenience for Long-term Benefits

Shedding loads during peak demand is important because it reduces the need for new power plants. To respond to high peak demand, utilities build very capital-intensive power plants and lines.

Peak demand happens just a few times a year, so those assets run at a mere fraction of their capacity. Electric users pay for this idle capacity through the prices they pay for electricity. According to the Demand Response Smart Grid Coalition, 10%-20% of electricity costs in the United States are due to peak demand during only 100 hours of the year. DR is a way for utilities to reduce the need for large capital expenditures, and thus keep rates lower overall; however, there is a economic limit to such reductions because consumers lose the productive or convenience value of the electricity not consumed. Thus, it is misleading to only look at the cost savings that demand response can produce without also considering what the consumer gives up in the process.

Importance for the Operation of Electricity Markets

It is estimated that a 5% lowering of demand would have resulted in a 50% price reduction during the peak hours of the California electricity crisis in 2000/2001. With consumers facing peak pricing and reducing their demand, the market should become more resilient to intentional withdrawal of offers from the supply side.

Residential and commercial electricity use often vary drastically during the day, and demand response attempts to reduce the variability based on pricing signals. There are three underlying tenets to these programs:

1. Unused electrical production facilities represent a less efficient use of capital (little revenue is earned when not operating).
2. Electric systems and grids typically scale total potential production to meet projected peak demand (with sufficient spare capacity to deal with unanticipated events).
3. By "smoothing" demand to reduce peaks, less investment in operational reserve will be required, and existing facilities will operate more frequently.

In addition, significant peaks may only occur rarely, such as two or three times per year, requiring significant capital investments to meet infrequent events.

Initiative of the US Energy Policy Act of 2005

The United States Energy Policy Act of 2005 has mandated the Secretary of Energy to submit to the US Congress "a report that identifies and quantifies the national benefits of demand response and makes a recommendation on achieving specific levels of such benefits by January 1, 2007." Such a report was published in February 2006.

The report estimates that in 2004 potential demand response capability equalled about 20,500 megawatts (MW), 3% of total U.S. peak demand, while actual delivered peak demand reduction was about 9,000 MW (1.3% of peak), leaving ample margin for improvement. It is further estimated that load management capability has fallen by 32% since 1996. Factors affecting this trend include fewer utilities offering load management services, declining enrollment in existing programs, the changing role and responsibility of utilities, and changing supply/demand balance.

To encourage the use and implementation of demand response in the United States, the Federal Energy Regulatory Commission (FERC) issued Order No. 745 in March 2011, which requires a certain level of compensation for providers of economic demand response that participate in wholesale power markets. The order is highly controversial and has been opposed by a number of energy economists, including Professor William, W. Hogan at Harvard University's Kennedy School. Professor Hogan asserts that the order overcompensates providers of demand response, thereby encouraging the curtailment of electricity whose economic value exceeds the cost of producing it. Professor Hogan further asserts that Order No. 745 is anticompetitive and amounts to "...an application of regulatory authority to enforce a buyer's cartel." Several affected parties, including the State of California, have indicated that they may challenge in federal court the legality of Order 745.

Demand Reduction and the Use of Diesel Generators in the UK National Grid

As of December 2009 UK National Grid had 2369 MW contracted to provide Demand Response, known as STOR, the demand side provides 839 MW (35%) from 89 sites. Of this 839 MW approximately 750 MW is back-up generation with the remaining being load reduction.

Feed-in Tariff

A feed-in tariff (FIT, standard offer contract advanced renewable tariff or renewable energy payments) is a policy mechanism designed to accelerate investment in renewable energy technologies. It achieves this by offering long-term contracts to renewable energy producers, typically based on the cost of generation of each technology. Technologies such as wind power, for instance, are awarded a lower per-kWh price, while technologies such as solar PV and tidal power are offered a higher price, reflecting higher costs.

In addition, feed-in tariffs often include "tariff degression", a mechanism according to which the price (or tariff) ratchets down over time. This is done in order to track and encourage technological cost reductions. The goal of feed-in tariffs is to offer cost-based compensation to renewable energy producers, providing the price certainty and long-term contracts that help finance renewable energy investments.

Description

FITs typically include three key provisions:

- guaranteed grid access
- long-term contracts for the electricity produced
- purchase prices based on the cost of generation.

Under a feed-in tariff, eligible renewable electricity generators (which can include homeowners, business owners, farmers, as well as private investors) are paid a cost-based price for the renewable electricity they produce. This enables a diversity of technologies (wind, solar, biogas, etc.) to be developed, providing investors a reasonable return on their investments. This principle was first explained in Germany's 2000 RES Act:

> *"The compensation rates...have been determined by means of scientific studies, subject to the provision that the rates identified should make it possible for an installation – when managed efficiently – to be operated cost-effectively, based on the use of state-of-the-art technology and depending on the renewable energy sources naturally available in a given geographical environment."*

As a result, the tariff (or rate) may differ to enable various technologies to be profitably developed. This can include different tariffs for projects in different locations (e.g. rooftop or ground-mounted for solar PV projects), of different sizes (residential or commercial scale), and sometimes, for different geographic regions. The tariffs are typically designed to ratchet downward over time to both track, and encourage, technological change.

FITs typically offer a guaranteed purchase agreement for electricity generated from renewable energy sources. These agreements are generally framed within long-term (15–25 year) contracts.

The fact that the payment levels are performance-based puts the incentive on producers to maximise the overall output and efficiency of their project.

As of 2011, feed-in tariff policies have been enacted in over 50 countries, including in Algeria, Australia, Austria, Belgium, Brazil, Canada, China, Cyprus, the Czech Republic, Denmark, Estonia, France, Germany, Greece, Hungary, Iran, Republic of Ireland, Israel, Italy, Kenya, the Republic of Korea, Lithuania, Luxembourg, the Netherlands, Portugal, South Africa, Spain, Switzerland, Tanzania, Thailand, Turkey

In 2008, a detailed analysis by the European Commission concluded that "well-adapted feed-in tariff regimes are generally the most efficient and effective support schemes for promoting renewable electricity". This conclusion has been supported by a number of recent analyses, including by the International Energy Agency, the European Federation for Renewable Energy, as well as by Deutsche Bank.

History

The United States

The first form of feed-in tariff was implemented in the US in 1978 under President Jimmy Carter, who signed the National Energy Act (NEA). This Act included five separate Acts, one of which was the Public Utility Regulatory Policies Act (PURPA). The purpose of the National Energy Act was to encourage energy conservation and the development of new energy resources, including renewables such as wind, solar and geothermal power.

Within PURPA was a provision that required utilities to purchase electricity generated from qualifying independent power producers at rates not to exceed their avoided cost. Avoided costs were designed to reflect the cost that a utility would incur to provide that same electrical generation. Different interpretations of PURPA prevailed in the 1980s: some utilities and state utility commissions interpreted avoided costs narrowly to mean avoided fuel costs, while others chose to define "avoided costs" as the "avoided long-run marginal cost" of generation. The long-run costs referred to the anticipated cost of electricity in the years ahead. This last approach was adopted by California in its Standard Offer Contract No. 4. Another provision included in the PURPA law was that utilities were prevented from owning more than 50% of projects, a clause that was introduced to encourage new participants to enter the electricity generation industry.

In order to comply with PURPA, certain states began offering Standard Offer Contracts to renewable power producers. California's Public Utility Commission established a number of Standard Offer Contracts, including Standard Offer No.4 (SO4), which made use of fixed prices, based on the expected long-run cost of generation. The long-run estimates of electricity costs were based on the belief (widely held at the time) that oil and gas prices would continue to increase. This led to an escalating schedule of fixed purchase prices, designed to reflect the long-run avoided costs of new electrical generation. By 1992, private power producers had installed approximately 1,700 MW of wind capacity in California, some of which is still in service today. The adoption of PURPA also led to significant amounts of renewable energy generation in states such as Florida, and Maine.

This notwithstanding, PURPA continues to have negative connotations in the U.S. electricity industry. When oil and gas prices plummeted in the late 1980s, the Standard Offer Contracts that were signed to encourage new renewable energy development seemed high by comparison. As a result, PURPA contracts came to be seen as an expensive burden on electricity ratepayers, a connotation that remains to this day.

Another source of opposition to PURPA stems from the fact that it was designed to encourage non-utility generation. This was interpreted as a threat by many large utilities at the time, particularly those operating in monopoly environments. As a result of its encouragement of non-utility generation, PURPA has been interpreted as an important step toward increasing competition in the U.S. electricity industry by some.

Developments in Europe

In 1990, Germany adopted its "Stromeinspeisungsgesetz" (StrEG), or its "Law on Feeding Electricity into the Grid" The StrEG required utilities to purchase electricity generated from renewable energy sources at prices that were determined as a percentage of the prevailing retail price of electricity. The percentage offered to solar and wind power was set at 90% of the residential electricity price, while other technologies such as hydro power and biomass sources were offered percentages ranging from 65–80%. A project cap of 5 MW was included

While Germany's StrEG (Feed-in Law) was insufficient to encourage costlier sources of generation such as solar photovoltaics, it proved relatively effective at encouraging lower-cost technologies such as wind power, leading to the deployment of 4 400 MW of new wind capacity between 1991 and 1999, representing approximately one third of the global capacity at the time.

An additional challenge that was addressed by Germany's Feed-in Law was the right to interconnect to the utility's grid. The StrEG stipulated that renewable electricity producers would be guaranteed grid access. Similar percentage-based feed-in laws were adopted in Spain, as well as in Denmark in the 1990s.

Germany's Renewable Energy Sources Act

Germany's Feed-in Law underwent a major restructuring in the year 2000, being re-framed as the Act on Granting Priority to Renewable Energy Sources ("Erneuerbare Energien Gesetz", German Renewable Energy Act). In its new form, it has proved to be the world's most effective policy framework at accelerating the deployment of renewable energy technologies.

Germany's new feed-in tariff made a number of important changes to its previous policy:

1. first, the purchase prices were methodologically based on the cost of generation from renewable energy sources. This led to different prices for wind power, solar power, biomass and biogas sources, and geothermal energy, as well as different prices for projects of different sizes, to account for economies of scale;
2. purchase guarantees were extended for a period of 20 years;
3. utilities were now allowed to participate; and finally
4. the rates offered were designed to decline annually based on expected cost reductions, in a mechanism known as "tariff degression".

Since it has been the most successful, the German policy (amended in 2004 and 2008) often provides the benchmark against which other feed-in tariff policies are considered.

Following the German approach, a number of countries have begun adopting feed-in tariff policies. These long-term contracts for electricity are typically offered in a non-discriminatory manner to all producers of electricity generated from renewable energy sources. Because the purchase prices are based on the cost of generation, investors are assured that efficiently operated projects will yield a reasonable rate of return.

This principle was explained in Germany's 2000 RES Act:

> *"The compensation rates...have been determined by means of scientific studies, subject to the proviso that the rates identified should make it possible for an installation – when managed efficiently – to be operated cost-effectively, based on the use of state-of-the-art*

technology and depending on the renewable energy sources naturally available in a given geographical environment."

Feed-in tariff policies typically target a rate of return ranging from 5–10%.

Feed-In Tariffs (REFIT) have been associated with a large growth in solar power in Spain, Germany and wind power in Denmark.

Effects on Electricity Rates

Cost increases of approximately four Euros per month per household are recorded in Germany. A number of analyses have shown that these price increases, however, can be partially offset by the price-dampening effect that large amounts of lower cost renewable energy sources (such as wind power) can have on spot market prices. This is attributed in large part to what is called the merit order effect. This has led to electricity price reductions in Spain, Denmark, and Germany. Cost impacts will depend to a significant extent on the overall mix of renewable energy sources encouraged by the policy. Due to the currently higher cost of certain technologies such as solar PV, the deployment of large amounts of these technologies would lead to proportionally larger rate payer impacts. In Germany, the EEG surcharge rose from 2.047 ct/kWh in 2010 to 3.53 ct/kWh in 2011 and 3.592 ct/kWh in 2012. With the EEG surcharge (EEG-Umlage, also called EEG reallocation charge), consumers pay for the difference between the guaranteed feed-in tariffs and the sale of the renewable energy at the energy exchange by the transmission system operators. The increase from 2010 to 2011 mainly stemmed from the rise in PV capacity in 2010 due to favourable solar feed-in tariffs.

Grid Parity

Some have argued that feed-in tariffs can be used to accelerate the pace at which renewable energy technologies become cost-competitive with electricity provided from the grid. The rapid deployment of renewable energy under feed-in tariffs seen in countries like Germany, Denmark and Spain has undoubtedly contributed to reducing technology costs, and hence, in accelerating this trend. For instance, wind and solar technology costs have decreased dramatically since the 1960s and 1970s, as the technologies have become more widespread, manufacturing processes have improved, innovations have been incorporated, and gains have been harnessed from economies of scale.

While it is true that large scale deployment of renewable energy technologies contributes to advancing toward grid parity, grid parity itself is a moving target, both in time (i.e. during the course of the day and over the course of years) and in space (i.e. geographically). The price of electricity from the grid varies widely from high-cost jurisdictions such as Hawaii and California, to lower-cost jurisdictions such as Wyoming, and Idaho. Similarly, due to their dependence on diesel generators, islands typically have higher electricity costs than on the mainland. In jurisdictions with time-of-use pricing the electricity price changes over the course of the day, rising during high-demand hours (e.g. 11AM – 8 PM) and declining during low-demand hours.

In certain jurisdictions, wind power, landfill gas, and certain forms of biomass generation are already lower-cost (on a per-kWh basis) than electricity provided from the grid. In fact, "grid parity" has already been obtained in certain jurisdictions that continue to use feed-in tariffs (e.g. the generation cost from landfill gas systems in Germany are currently lower than the average electricity spot market price) And in remote areas, electricity from solar photovoltaics can be cheaper than building new distribution lines to connect up to the main transmission grid. There are other reasons to retain FIT laws even as grid parity approaches, because there are other important aspects of FIT laws which should be retained even if remuneration drops to 1:1 equivalence with the retail electricity price. These include the obligation to purchase, the obligation to connect and the long term payment guarantee associated with FIT laws.

Policy Alternatives and Complements

Schemes such as quota-based mechanisms (Renewable Portfolio Standards) and subsidies create limited protected markets for renewable energy. The supply of renewable energy is achieved by obliging suppliers to deliver to consumers a portion of their electricity from renewable energy sources. To do this, they collect green electricity certificates. Hence, a market is created in green electricity certificates which, according to the theory, generates downward pressure on the prices paid to renewable energy developers. This is based on the theory of perfect competition where there is a multiplicity of buyers and sellers in a market where no single buyer or seller has a big enough market share to have a significant influence on prices. Although, in practice, markets are very rarely perfectly competitive, the assumption is still that a relatively competitive market will produce a more efficient use of resources compared to a system where prices are set by analysts.

The fundamental problem with the quota scheme is that there is no long-term certainty. When a quota is set either for a period of time or for a quantity of power, once that goal is reached then there is nothing to keep the green power producers from becoming uneconomic in the face of power produced from coal-fired power stations and hence collapsing as businesses. This inevitability with the quota method means that there is reluctance on behalf of investors to get involved in the first place. Those that do get involved are short-term speculators rather than long-term entrepreneurs and so instability is inherent in this system. Quota systems favour large, vertically integrated generators and multinational electric utilities, and are more difficult to design and implement than a price system.

Territories

Feed-in tariff laws were in place in 46 jurisdictions across the world by 2007. Research for information on feed in tariffs is difficult to find and typically extensive research is required to find it. Information about solar feed in tariffs may be found in a consolidated form, however not all of the countries are listed in this source.

Algeria

To cover the additional costs of producing electricity from renewables and for the costs of diversification, producers of electricity from renewables will get a bonus for each kWh produced, marketed or consumed. For electricity generated from solar or radiant heat only, the bonus is 300% of the price per kWh of electricity produced by the market operator defined by Law 02-01 of 22 Dhu El Kaada 1422 corresponding to 5 February 2002 above, and this when the minimum contribution of solar energy represents 25% of all primary energy. For electricity generated from facilities using solar thermal systems solar-gas hybrid, the bonus is 200% of the price per kWh.

For contributions of solar energy below 25%, said the bonus is paid in the following conditions:

Solar Share	*Bonus*
>25%	200%
20% to 25%	180%
15% to 20%	160%
10% to 15%	140%
5% to 10%	100%
0 to 5%	0

The price of electricity is fixed by the CREG (Gas and Electricity Regulatory Commission). According to the last decision which fixed it, the consumer pays his electricity as below:

- 1.77 DZD/kWh for a consumption which is lower than 41.6 kWh/month.
- 4.17 DZD/kWh for a consumption which is higher than 41.6 kWh/month.

The other consumers (Industry, agriculture...etc.), they pay their electricity 4.17 DZD/kWh.

Furthermore the feed-in tariff provides with bonuses for electricity generated by cogeneration. The bonus will be 160% taking into account a thermal energy use by 20% of all primary energy used. The bonuses for solar generated electricity and cogeneration are cumulative. Remuneration of the generated electricity will be guaranteed over the whole plant lifetime.

Australia

Feed-in tariffs had been introduced by the States in 2008 in South Australia and Queensland, 2009 in the Australian Capital Territory and Victoria and 2010 in New South Wales, Tasmania and Western Australia. In the Northern Territory there are only local feed-in tariff schemes. A uniform federal scheme to supersede all State schemes has been proposed by Tasmanian Greens Senator Christine Milne, but not enacted. By mid 2011, FIT in NSW and ACT had been closed to new generators, as the installed capacity cap had been reached. In NSW, both the FiT and the cap were cut, due to the overly generous original settings.

The new conservative Victorian government has replaced the original FiT with a less generous transitional FiT of 25c per kilowatt hour for any power generated excess to the generator's usage, pending the outcome of an inquiry by the Victorian Competition and Efficiency Commission.

Canada

Ontario introduced a feed-in tariff in 2006, and revised it in 2009, which in a draft proposal increases from 42¢/kWh to 80.2¢/kWh for micro-scale (≤10 kW) grid-tied photovoltaic projects. Ontario's FIT program also includes a tariff schedule for larger projects up to and including 10MW solar farms at a reduced rate. As of April 2010, several hundred projects have been approved, including 184 large scale projects, worth $8 billion all together.

China

As of August 2011 a national feed-in tariff for solar projects was issued, and is about us$0.15 per kWh.

"China has set a fixed feed-in tariff for new onshore wind power plants in a move that will help struggling project operators to realise profits. The National Development and Reform Commission (NDRC), the country's economic planning agency, announced at the weekend four categories of onshore wind projects, which according to region will be able to apply for the tariffs. Areas with better wind resources will have lower feed-in tariffs, while those with lower outputs will be able to access more generous tariffs.

The tariffs per kilowatt hour are set at 0.51 yuan (US 0.075, GBP 0.05), 0.54 yuan, 0.58 yuan and 0.61 yuan. These represent a significant premium on the average rate of 0.34 yuan per kilowatt hour paid to coal-fired electricity generators.".

Czech Republic

Czech Republic introduced a feed-in tariff by act of law no. 180/ 2005 for wide range of renewable sources in 2005. Tariff is guaranted for 15–30 years (depends on source). Supported sources are small hydropower (up to 10 MW), biomass, biogas, wind and photovoltaics. As of 2010 highest tariff is 12.25 CZK/kWh for small photovoltaic.

France

The administrative procedure for the ground-mounted PV systems market segment was significantly modified in late 2009. The distinction between the segments is essentially based on the capacity of the PV installation, which determines the complexity of the administrative process. Differences between administrative procedures on regional level have not been detected. Since the decree of 4 March 2011, the feed-in tariff (FiT) is only attractive for installations less than 100 kWp in size. Above 100 kWp, the FiT is established at 0,117 €/kWh (until 30th september 2011), thus severely limiting the profitability of the installation. A call for tenders for PV projects above 250 kWp has been launched on September 15th 2011. Through this call, PV projects are analysed on multiple criteria, including the FiT requested by the applicant. The results of this call for tenders will be knew only at the end of March 2012.

Medium sized PV installations on roofotops up to 100 kWp in size are essentially supported by the feed-in tariff (FiT). The main obstacles for the development of the PV system are the grid connection procedure and the vagueness of certain regulations. Differences between administrative procedures at regional level have not been detected. Since the decree of 4 March 2011, FIT is attractive only for installations up to 100 kWp in size.

Above 100 kWp, it is limited to 11,688cts€/kWh (until the next revie on 30th September 2011) this severely limiting the profitability of the PV installation. A call for tenders for rooftop PV projects above 100 kWp has been launched on August 1st 2011. This call will evaluate PV projects on the basis of the FIT requested by the applicant.

Germany

First introduced in 2000, the Erneuerbare-Energien-Gesetz (EEG) law is reviewed on a regular basis and the 2012 version is currently in force. Its predecessor was the 1991 "Stromeinspeisegesetz". As of May 2008, the cost of the program added about €1.01 (USD1.69) to each monthly residential electric bill.

In 2012 the additional costs have risen to €0.03592/kWh. Since 2009, there are additional tariffs for electricity immediately consumed rather than supplied to the grid with increasing returns if more than 30% of overall production is consumed on-site. This is to incentivise a demand side management and help develop solutions to the intermittency of solar power.

Since 2009, there are additional tariffs for electricity immediately consumed rather than supplied to the grid with increasing returns if more than 30% of overall production is consumed on-site. This is to incentivise a demand side management and help develop solutions to the intermittency of solar power.

India

India's inaugurated its most ambitious solar power program to date on 9 January 2010. The Jawaharlal Nehru National Solar Mission (JNNSM) was officially announced by Prime Minister of India on 12 January 2010. This program aims to install 20,000 MW of solar power by 2022.

The first phase of this program aims to install 1000 MW by paying a tariff fixed by the Central Electricity Regulatory Commission (CERC) of India. While in spirit this is a feed in tariff, there are several conditions on project size and commissioning date. Tariff for solar PV projects is fixed at Rs. 17.90 (USD 0.397/kWh). Tariff for solar thermal projects is fixed Rs. 15.40 (USD 0.342/kWh). Tariff will be reviewed periodically by the CERC.

Iran

(SANA), first introduced FiT in 2008 for purchasing renewable energy from investors. A price of 1300 Rials/kWh was set for renewable electricity. For 4 hours in the mid-night, the price is 900 Rials.

Republic of Ireland

From 2006 to 2011 the sitting Ministers for Communications Energy and Natural Resources (Noel Dempsey, Eamon Ryan) have at various times announced Feed in Tariffs for renewable electricity in Ireland. However they have never actually put feed in tariffs in place. This has seriously hindered the growth of renewable electricity in Ireland. Currently (Q2:2011) REFIT is being negotiated between DCENR and the EU, the industry is hopeful that REFIT will be available, however much investment funds have been lost to oversees projects at this stage. Industry associations (such as the Irish BioEnergy Association) are strongly lobbying government and government departments to have feed in tariffs available to the renewable energy sector.

Israel

On 2 June 2008, the Israeli Public Utility Authority approved a *feed-in tariff* for solar plants. The tariff is limited to a total installation of 50MW during 7 years, whichever is reached first, with a maximum of 15 kWp installation for residential and a maximum of 50kWp for commercial. Bank Hapoalim offered 10 year loans for the installation of solar panels. The National Infrastructures Ministry announced that it would expand the *feed-in tariff* scheme to include medium-sized solar-power stations ranging from 50 kilowatts to 5 megawatts. The new tariff scheme caused solar company Sunday Solar Energy to announce that it would invest $133 million to install photovoltaic solar arrays on kibbutzim, which are social communities that divide revenues amongst their members.

The Netherlands

The Netherlands has no FIT yet. However, the Dutch Cabinet agreed on 27 March 2009 to implement some parts of a feed-in tariff during a renegotiation of the government agenda in response to the global financial crisis. The proposed regulation may adjust the quota incentive system. As of the summer of 2009, The Netherlands has a subsidy system. The subsidy budget has a quota for diverse types of energy, at several tens of million euro. The budget for wind is hardly used, because the tariffs are too low. The 2009 budget for Wind on

Land was 900 MW (incl unused 400 MW from 2008); only 2.5 MW was used. Dutch utilities have no obligation to buy energy from windparks. The tariffs change yearly, to compensate the change in market prices. This creates uncertain investment conditions. The current subsidy system was introduced in 2008. The previous subsidy scheme stopped in 2005 because it became "too expensive". In 2009, Dutch windparks are still being built with grants from the old scheme. The old and new subsidy scheme is fed from the general budget.

The most probable change towards the FIT is that the subsidy budget will be fed with a small increase of the kWh price. The quota remain. End of November 2009 the Dutch Government spoke out that they now finally will also follow the Germans with their FiT to make investments in Sustainable Energy for investors much more interesting. Details will follow shortly after November 2009.

South Africa

"South Africa's National Energy Regulator (NERSA) announced 31 March 2009 the introduction of a system of feed-in tariffs designed to produce 10 TWh of electricity per year by 2013. The feed-in tariffs announced were substantially higher than those in NERSA's original proposal. The tariffs, differentiated by technology, will be paid for a period of 20 years.

NERSA said in its release that the tariffs were based, as in most European countries, on the cost of generation plus a reasonable profit. The tariffs for wind energy and concentrating solar power are among the most attractive worldwide.

The tariff for wind energy, 1.25 ZAR/kWh (€0.104/kWh, $0.14 USD/kWh, $0.17 CAD/kWh) is greater than that offered in Germany (€0.092/kWh) and more than that proposed in Ontario, Canada ($0.135 CAD/kWh). The tariff for concentrating solar, 2.10 ZAR/kWh (€0.175/kWh), is less than that in Spain (€0.278/kWh), but offers great promise in the bright sunlight of South Africa. NERSA's revised program followed extensive public consultation.

Stefan Gsänger, Secretary General of the World Wind Energy Association said in a release that "South Africa is the first African country to introduce a feed-in tariff for wind energy. Many small and big investors will now be able to contribute to the take-off of the wind industry in the country. Such decentralised investment will enable South Africa to overcome its current energy crisis. It will also help many South African communities to invest in wind farms and generate electricity, new jobs and new income. We are especially pleased as this decision comes shortly after the first North American feed-in law has been proposed by the Government of the Canadian Province of Ontario".

However, the feed-in tariff was abandoned before being promulgated in favour of a competitive bidding process launched on August 3, 2011. Under this bidding process, the South African government plans to procure 3,750MW of renewable energy: 1,850MW of onshore wind, 1,450MW of solar PV, 200MW of CSP, 75MW of small hydro, 25MW of landfill gas, 12.5MW of biogas, 12.5MW of biomass, and 100MW of small projects. The bidding process comprises two steps:

- Qualification phase. Projects are assessed based on structure of the project, legal, land acquisition and use, financial, environmental consent, technical, economic development and bid guarantee
- Evaluation phase. Compliant bids are then evaluated based on: (1) price relative to a ceiling provided in bid documentation, accounting for 70% of the decision, and (2) economic development, accounting for 30% of the decision.

The first round of bids is due on November 4, 2011. The SA government is expected to announce preferred bidders before COP17 in December. PPA's are expected to be in place by June 2012. Projects should commission by June 2014, except CSP projects which are expected by June 2015.

Spain

Spanish feed-in legislation was set by Royal decree 1578/2008 (Real Decreto 1578/2008), for photovoltaic installations, and Royal decree 661/2007 for other renewable technologies injecting electricity to the public grid. Originally under the 661/2007, photovoltaic feed-in tariffs have been developed under a separate specific law frame due to the rapid growth experienced by this technology since release of the original scheme.

The decree 1578/2008 categorises installations in two main groups with differentiated tariffs:

I. Building Integrated installations; with 34c€/kWh in systems up to 20 kW of nominal power, and for systems above 20 kW with a limit of nominal power of 2MW tariff of 31c€/kWh.

II. Non integrated installations; 32c€/kWh for systems up to 10MW of nominal power.

For other technologies decree 661/2007 sets up:

Energy Source	*Feed-in Tariff*
Cogeneration systems	maximum FiT of 13.29c€/kWh during lifetime of system.
Solar thermoelectric	26.94 c€/kWh for the first 25 years
Wind systems	up to 7.32 c€/kWh for the first 20 years
Geothermal, wave, tidal and sea-thermal	6.89 c€/kWh for the first 20 years
Hydroelectric	7.8 c€/kWh for the first 25 years
Biomass and biogas	up to 13.06 c€/kWh for the first 15 years
Waste combustion	up to 12.57 c€/kWh for the first 15 years

In January 2012 the Spanish government froze all subsidies to renewable energy.

Switzerland

Switzerland introduced the so called "Cost-covering remuneration for feed-in to the electricity grid (CRF)" on 1 May 2008. In German, it is called "KEV", Kostendeckende Einspeise-Vergütung.

The CRF/KEV applies to hydropower (up to 10 megawatts), photovoltaics, wind energy, geothermal energy, biomass and waste material from biomass and will be applicable for a period of between 20 and 25 years, depending on the technology. The implementation is done through the national grid operator SWISSGRID. While high by appearance, the KEV tarif has been almost inconsequential, as the total amount of "extra" cost to the system has been severly limited and capped by law. Since about 2009, no more projects could be financed through the Swiss federal KEV scheme. About 15'000 projects are waiting for allocation of monies from the KEV scheme. If all those projects would be passed, Switzerland could immediately turn-off all its nuclear power plants, which currently supply 40% of the power. Nevertheless, in 2011 after Fukushima, some local power companies, mostly owned by villages and cantons/provinces, have selectively started offering either the full KEV tarif as decided and calculated by the Swiss federal department of Energy BfE, or some part of that tarif, thereby creating a mini-boom in renewable energy.

As of March 2012 the KEV-FIT for Solar PV has been lowered several times to CHF 0.30-0.40/kWh (USD 0.33-0.44/kWh) depending on size, but is still higher than in Germany and most of the rest of the world. Politically there are many initiatives to broaden the approach of KEV. There is also a political discussion suggesting that a subsidy through a FIT mechanism is really only needed as long as the "external costs" of traditional power production are not included in the price of traditional nuclear and fossil electric power. E.g. the cost of nuclear waste disposal over a million years, or the cost of insuring against having a nuclear plant destroying most of Switzerland, or the true cost of CO2 which is placed at about EUR 200.- / tCO2, instead of the 2011-EU-ETS-pseudo-market-price of around EUR 5-10 / tCO2 have never been added to power prices. Once that was to take place and power prices would be requested to reflect full costs including all external costs, most renewable energy solutions would become instantaneous cost-leaders, eliminating the need for a FIT mechanism. Unfortunately, big-industry-lobbies in Switzerland are typically blocking the adoption of a true-cost mechanism for power. As a result, Switzerland's clean energy industry has fallen behind Germany, Italy, China and Spain in terms of clean energy competitiveness.

Thailand

In 2006, the Thai government enacted a feed-in tariff that provides an adder paid on top of utility avoided costs, differentiated by technology type and generator size, and guaranteed for 7–10 years. Solar receives the highest, 8 baht/kWh (about US cents 27/kWh). Large biomass projects receive the lowest at 0.3 baht/kWh (at about 1 US cent per kWh). Additional per-kWh subsidies are provided for projects that offset diesel use in remote areas. Under the FIT program, as of March 2010, 1364 MW of private sector RE was online with an additional 4104 MW in the pipeline with signed PPAs. Biomass makes up the bulk of this capacity: 1292 MW (online) and 2119 MW (PPA only). Solar electricity is second but rapidly catching up, with 78 MW online and signed PPAs for an additional 1759 MW.

Uganda

Uganda has launched a feed-in tariff in 2011. The Uganda Electricity Transmission Company Limited holds the transmission license in the country and was mandated by the Electricity Regulatory Authority to provide the following FIT for small-scale projects ranging from 0.5MW to 20MW.

Ukraine

Ukraine has introduced the law 'On feed-in tariff' as of 25 September 2008. The law guarantees grid access for renewable energy producers (small hydro up to 10 MW, wind, biomass, photovoltaic and geothermal). The feed-in tariffs for renewable power producers are set by the national regulator. As of July 2011 the following tariffs per kWh were applied: biomass – UAH 1.3446 (EUR 0.13), wind – UAH 1.23 (EUR 0.12), small hydro – UAH 0.8418 (EUR 0.08), solar-UAH 5.0509 (EUR 0.46). In case of significant fluctuations of the national currency against Euro the feed-in tariff is adjusted to reflect the changes.

United Kingdom

Feed-in tariffs in the United Kingdom were first announced in October 2008 by the UK Secretary of State for Energy and Climate Change, Ed Miliband, who announced that Britain would implement a scheme by 2010, in addition to its current renewable energy quota scheme (ROCS). In July 2009, he presented details of the scheme, which began in early April 2010.

Less than a year into the scheme, in March 2011 the new coalition government announced that support for large-scale photovoltaic installations (greater than 50 kW) would be cut. This was in response to European speculators lining up to establish huge solar farms in the West Country, which would have absorbed disproportionate amounts of the fund.

On 9 June 2011, DECC confirmed that Feed-in Tariffs cut for solar PV systems above 50 KW after 1st Aug, 2011. Many are disappointed with the decision of DECC, especially after long time discussion with public. Still some think positive about the new Feed-in Tariffs, from the report published by PV insights. It is believed that the total subsidies for solar PV industry is not changed, UK government tries to trim the tariffs for large systems in order to benefit smaller systems. The fast track review is based on the long term plan to reach an annual installation of 1.9GW in 2020. In October 2011 DECC announced dramatic cuts of around 55% to feed in tariff rates, with additional reductions for community or group schemes. The cuts were to be effective from 12 December 2011, with a consultation exercise to end on 23 December 2011.

This was successfully challenged in the high court by an application for judicial review, jointly made by environmental pressure group Friends of the Earth (FoE) and two solar companies-Solar century and HomeSun. The judgment, made by Mr Justice Mitting after a two-day court hearing, was hailed as a major victory by green campaigners and the solar industry. Lawyers for the Department of Energy and

Climate Change immediately moved to apply for permission to appeal the judge's ruling.

United States

In April 2009, 11 United States state legislatures were considering adopting a FiT as a complement to their renewable electricity mandates.

"While feed-in tariffs are most closely associated with solar photovoltaic panels, utilities managing the programs in Vermont and California will also pay a set price for electricity generated from other renewable sources, like wind.

California's Sacramento Municipal Utility District (SMUD) program is open to homeowners who are not participating in another program, called net metering, which allows anyone whose system is producing more electricity than they need to sell the excess back to the utility, thus reducing their electric bill. But once their bill falls to zero, the homeowner gets no more money from the system.

Jon Bertolino, a spokesman for SMUD, said that customers with land to spare had been asking whether, if they put up small solar farms, the utility would buy the excess electricity.

As long as they are not part of the net-metering program and not seeking the $1.90- to $2.20-per-watt ratepayer subsidy for their new panels under the state's "Million Solar Roofs" program, Mr. Bertolino said, small generators can sell their power to SMUD. The rates would depend on the time of day the power is generated, ranging from a low of 5 or 6 cents a kilowatt-hour to 30 cents on a hot summer afternoon; the size of eligible systems is capped at 5 megawatts (and the program overall has a 100-megawatt cap). The Vermont law caps the size of individual systems at 2.2 megawatts. Solar energy fetches a fixed price of 30 cents a kilowatt-hour, and other forms of renewables fetch lower rates".

California

The California Public Utilities Commission (CPUC) approved a feed-in tariff on 31 January 2008 that is effective immediately.

As of 1 January 2010 state laws have passed allowing homeowners who produce more than they use over the course of the year to opt in to be able to sell the excess power to the utility at a currently undisclosed rate (the rate decision has a 1 January 2011 deadline). Previously the homeowner would get no credit for over-production over the course of the year. In order to get the California Solar Initiative (CSI) rebate the customer is not allowed to install a system that inherently over-produces thereby encouraging efficiency measures to be installed after solar installation. This over-production credit is not available to certain municipal utility customers namely Los Angeles Water and Power.

Florida

In February 2009, city commissioners in Gainesville, Florida, approved the nation's first solar feed-in tariff. The program is open to small, medium and large projects, but is capped at 4 MW per year. As of 2011, Gainesville has increased solar generated electricity from 328 kW to 7,391 kW. Solar energy now makes up approximately 1.2% of peak load energy (610 mW) in Gainesville.

Hawaii

In September 2009 the Hawaii Public Utilities Commission required Hawaiian Electric Company (HECO & MECO & HELCO) to pay above-market prices for renewable energy fed into the electric grid. The policy offers project a set price and standard 20-year contract. The PUC plans to review the initial feed-in tariff two years after the program starts and every three years thereafter.

Table : Feed-in tariffs – Wind & Hydropower

Renewable Generator Type and Size	***FIT Rate (cents/kWh)***
Baseline FIT rate for any RPS-eligible technology under the max. size limit	13.8
Tier 1 On-Shore Wind < 20 kW	16.1
Tier 1 In-line Hydropower < 20 kW	21.3
Tier 2 On-Shore Wind 20-100/500 kW	13.8
Tier 2 In-line Hydropower 20-100/500 kW	18.9

Table : Feed-in tariffs - Photovoltaics (PV) & Concentrating Solar Power (CSP)

Renewable Generator Type and Size	***FIT Rate (cents/kWh)***	
	35% State tax credit	***24.5% Refundable tax credit***
Tier 1 PV < 20 kW	21.8	27.4
Tier 1 CSP < 20 kW	26.8	33.1
Tier 2 PV 20-100/500 kW	18.9	23.8
Tier 2 CSP 20-100/500 kW	25.4	27.5

The PUC's decision sets project size limits of five megawatts (MW) for the island of »Oahu and 2.72 MW for Maui and Hawaii island. The Commission's decision caps the total amount of feed-in tariff projects brought onto the electricity grid at 5% of the system peak on Oahu, Maui, and the Big Island for the first two years of the program. Tier 3 is still pending a Decision and Order based on the findings of the Reliability Standards Working Group (a "docket within the docket"). Tier 2 and 3 project size caps vary by island and by technology. Tier 2 includes larger systems that are less than or equal to: 100 kW-AC for on-shore wind and in-line hydropower on all islands; 100 kW-AC for PV and CSP on Lanai and Molokai; 250 kW-AC for PV on Maui and Hawaii; 500 kW-AC for CSP on Maui and Hawaii; and 500 kW-AC for PV and CSP on Oahu. Tier 3 covers systems larger than the Tier 2 caps.

Maine

In the 2009 Maine Legislature's session, a "Feed-In" Tariff bill, (LD 1450), introduced by Rep. Herbert Adams (D-Portland), was considered. It made it from the House to the Senate, where is was killed 21 May. Known as "An Act to Establish the Renewable Energy Resources Program" it was closely modelled on the German law. In June 2009 a pilot program was initiated however, and is available for projects up to 10MW in size.

Oregon

In June 2009, Oregon established a pilot solar volumetric incentive rate and payment program* with legislation. Under this incentive program, systems are paid for the kilowatt-hours (kWh) generated over a 15-year period, at a rate set at the time a system is initially enrolled in the program. The Oregon Public Utility Commission (PUC) was left with the discretion to establish rates and rules by 1 July 2010. The PUC established rates and rules in May 2010. This program must be offered by the three investor-owned utilities in Oregon and will be administered by the utilities, though the PUC will periodically re-evaluate rates. The program costs are recoverable in utility rates and utility-owned systems are not allowed to receive the incentive.

The pilot program installation cap is limited to an aggregate cap of 25 megawatts (MW) of solar photovoltaics (PV), with a maximum system size cap of 500 kilowatts (kW). The aggregate program cap will be spread equally over four years, with 6.25 MW of capacity being eligible to receive the incentive each year. The aggregate cap is divided up by utility based on 2008 retail sales revenue. PGE has a cap of 14.9 MW, Pacific Power has a cap of 9.8 MW, and Idaho Power has a cap of 0.4 MW. Idaho Power's program is limited to residential installations. Rates differ by system size and geographic zone. Small- and medium-scale systems participate in a program that is modelled after net metering and larger-scale systems participate in a competitive bidding program. Participating PV systems must be grid-connected, metered and meet all applicable codes and regulations. Systems must be "permanently installed" and must remain in service for the entire useful life.

Systems sized 100 kW or less can participate in the portion of the program based on net metering. Generating capacity of 20 MW of the aggregate cap is reserved for the net metering portion, with 12 MW available for residential and 8 MW available for small commercial systems. These residential and small commercial systems are paid for the amount of electricity generated, up to the amount of electricity consumed. In essence, customers are paid for the amount of utility electric load consumption that is offset by on-site solar photovoltaic generation. Unlike typical feed-in tariffs, customers can consume the electricity generated on-site and receive a production incentive – or a volumetric incentive payment – for the amount of electricity generated and consumed. To remove a perverse incentive to increase electricity consumption to receive a greater payment, the system must be appropriately sized to meet average electricity consumption. The rates are determined by the PUC and are based on annual system cost and annual energy output, differentiated by geographic zones. The cost estimates are based on installation data from Energy Trust of Oregon. The volumetric incentive rates are not the actual rates paid to the customer-generator; the actual rate paid is the volumetric incentive rate minus the retail rate. The volumetric incentive rates will be re-evaluated every six months and may be adjusted if necessary.

Vermont

Vermont adopted feed-in tariffs on 27 May 2009. Generators must possess a capacity of no more than 2.2 MW.

Puerto Rico

The net metering program pays the energy fed back to the grid at the same rate it is charged. The rate varies monthly around 23 cents per kilowatt. It is not actually paid but credited to the next month if the production exceeds consumption. At the end of the fiscal year (June) the excess if any is paid at a fixed 10 cents per KW of which 25% is retained for public schools. Then the cycle starts again.

To participate in the program a yearly insurance, means for disconnecting the system accessible outside of the building and specific brands of equipment dictated by the government are required. The Puerto Rico Electric Power Authority (PREPA), a government agency, is the sole provider of energy in the island.

Electrical Grid

An electrical grid is an interconnected network for delivering electricity from suppliers to consumers. It consists of three main components:

1. generating plants that produce electricity from combustible fuels (coal, natural gas, biomass) or non-combustible fuels (wind, solar, nuclear, hydro power);
2. transmission lines that carry electricity from power plants to demand centres; and
3. transformers that reduce voltage so distribution lines carry power for final delivery.

In the power industry, *electrical grid* is a term used for an electricity network which includes the following three distinct operations:

1. Electricity generation-Generating plants are usually located near a source of water, and away from heavily populated areas. They are usually quite large in order to take advantage of the economies of scale. The electric power which is generated is stepped up to a higher voltage-at which it connects to the transmission network.
2. Electric power transmission-The transmission network will move (wheel) the power long distances-often across state lines, and sometimes across international boundaries until it reaches its wholesale customer (usually the company that owns the local distribution network).
3. Electricity distribution-Upon arrival at the substation, the power will be stepped down in voltage—from a transmission level voltage to a distribution level voltage. As it exits the substation, it enters the distribution wiring. Finally, upon arrival at the service location, the power is stepped down again from the distribution voltage to the required service voltage(s).

Term

The term *grid* usually refers to a network, and should not be taken to imply a particular physical layout or breadth. Grid may also be used to refer to an entire continent's electrical network, a regional transmission network or may be used to describe a subnetwork such as a local utility's transmission grid or distribution grid.

History

Since its inception in the Industrial Age, the electrical grid has evolved from an insular system that serviced a particular geographic area to a wider, expansive network that incorporated multiple areas. At one point, all energy was produced near the device or service requiring that energy. In the early 19th century, electricity was a novel invention that competed with steam, hydraulics, direct heating and cooling, light, and most notably gas. During this period, gas production and delivery had become the first centralised element in the modern energy industry.

It was first produced on customer's premises but later evolved into large gasifiers that enjoyed economies of scale. Virtually every city in the U.S. and Europe had *town gas* piped through their municipalities as it was a dominant form of household energy use. By the mid-19th century, electric arc lighting soon became advantageous compared to volatile gas lamps since gas lamps produced poor light, tremendous wasted heat which made rooms hot and smoky, and noxious elements in the form of hydrogen and carbon monoxide. Modelling after the gas lighting industry, Thomas Edison invented the first electric utility system which supplied energy through virtual mains to light filtration as opposed to gas burners. With this, electric utilities also took advantage of economies of scale and moved to centralised power generation, distribution, and system management.

During the 20th century, institutional arrangement of electric utilities changed. At the beginning, electric utilities were isolated systems without connection to other utilities and serviced a specific service territory. In the 1920s, utilities joined together establishing a wider utility grid as joint-operations saw the benefits of sharing peak load coverage and backup power. Also, electric utilities were easily financed by Wall Street private investors who backed many of their ventures. In 1934, with the passage of the Public Utility Holding Company Act (USA), electric utilities were recognised as public goods of importance along with gas, water, and telephone companies and thereby were given outlined restrictions and regulatory oversight of their operations.

This ushered in the *Golden Age of Regulation* for more than 60 years. However, with the successful deregulation of airlines and telecommunication industries in late 1970s, the Energy Policy Act (EPAct) of 1992 advocated deregulation of electric utilities by creating wholesale electric markets. It required transmission line owners to allow electric generation companies *open access* to their network.

Deregulation

With deregulation, a more complex environment occurred as opposed to the traditional vertically-integrated monopoly that oversees the entire grid's operations. Newer participants entered the market including Independent Power Providers (IPPs) who decided and constructed the new facility; Transmission Companies (TRANSCOs) who constructed and owned the transmission equipment; retailers who signed up end-use customers, procured their electric service, and billed them; integrated energy companies (combined IPPs and retailers); and Independent System Operation (ISO) who managed the grid being indifferent to market outcomes. Also, day-to-day to long term operations altered.

Infrastructure additions which were long-term planning now became an investment analysis with IPPs that decided construction of a new power plant under economic considerations (i.e., taxes, labour and material costs) and ability to obtain financing. Load and supply management that fell under mid-term planning became risk management as private utilities had to manage a portfolio of end customers and assets with the company's risk preference. Day-ahead scheduling and real time grid management in the short-term planning which involves forecasting demand and dispatch schedule became asset management as power plants and grid equipment were assets to be scheduled and dispatched. Here, the ISO sets dispatch schedule at the market clearing price where the supply bids of generating units equilibriated with demand bids of retailers.

Many engineers argue the unfortunate disadvantages that stem from deregulation. Where under regulated monopolies, long distance energy lines were used for emergencies as backup in case of generation outages, now, particularly in North America, the majority of domestic generation is sold over ever-increasing distances on the wholesale market before delivery to customers. Consequently, the power grid witnesses fluctuating power flows that impact system stability and reliability. To reduce system failure, the power flow of a transmission line must operate below the transmission line's capacity. Yet now, companies are continually operating near capacity. Additionally, as utilities exchange power to other utilities, power flows along all paths of connection. Therefore, any change in one point of generation and transmission affects the load on all other points. Oftentimes, this is unanticipated and uncontrolled. Usually, a longer line's capacity is less than a shorter line's capacity.

If not, power-supply instability occurs resulting in transmission lines that break or sag. Such phase and voltage fluctuations cause system interruptions as witnessed in the Northeast Blackout of 1965 (which involved a circuit breaker to trip) and 2003 (which involved a sagging line on a tree that rippled in magnitude). Furthermore, IPPs add new generating units at random locations determined by economics that extend the distance to main consuming areas adversely affecting power supply. Also, utilities, because of competitive information needs, do not publicise needed data to predict and react to system stress such as with energy flows and blackout statistics. Overall, the economics of the electrical grid do not align sufficiently with the physics of the grid. Experts advocate for fundamental changes to avoid serious consequences in the near future.

Features

Structure of Distribution Grids

The structure, or "topology" of a grid can vary considerably. The physical layout is often forced by what land is available and its geology. The logical topology can vary depending on the constraints of budget, requirements for system reliability, and the load and generation characteristics.

The cheapest and simplest topology for a distribution or transmission grid is a *radial* structure. This is a *tree* shape where power from a large supply radiates out into progressively lower voltage lines until the destination homes and businesses are reached.

Most transmission grids require the reliability that more complex *mesh networks* provide. If one were to imagine running redundant lines between limbs/branches of a tree that could be turned in case any particular limb of the tree were severed. The expense of mesh topologies restrict their application to transmission and medium voltage distribution grids. Redundancy allows line failures to occur and power is simply rerouted while workmen repair the damaged and deactivated line. Other topologies used are *looped* systems found in Europe and *tied ring* networks. n cities and towns of North America, the grid tends to follow the classic *radially fed* design. A substation receives its power from the transmission network, the power is stepped down with a transformer and sent to a bus from which feeders fan out in all directions across the countryside. These feeders carry three-phase power, and tend to follow the major streets near the substation. As the distance from the substation grows, the fanout continues as smaller laterals spread out to cover areas missed by the feeders. This tree-like structure grows outward from the substation, but for reliability reasons, usually contains at least one unused backup connection to a nearby substation. This connection can be enabled in case of an emergency, so that a portion of a substation's service territory can be alternatively fed by another substation.

Geography of Transmission Networks

A wide area synchronous grid or "interconnection" is a group of distribution areas all operating with alternating current (AC) frequencies synchronised (so that peaks occur at the same time). This allows transmission of AC power throughout the area, connecting a large number of electricity generators and consumers and potentially enabling more efficient electricity markets and redundant generation.

In a synchronous grid all the generators run not only at the same speed but also at the same phase, each generator maintained by a local governor that regulates the driving torque by controlling the steam supply to the turbine driving it. Generation and consumption must be balanced across the entire grid, because energy is consumed almost instantaneously as it is produced. Energy is stored in the immediate short term by the rotational kinetic energy of the generators.

A large failure in one part of the grid-unless quickly compensated for-can cause current to re-route itself to flow from the remaining generators to consumers over transmission lines of insufficient capacity, causing further failures. One downside to a widely connected grid is thus the possibility of cascading failure and widespread power outage. A central authority is usually designated to facilitate communication and develop protocols to maintain a stable grid. For example, the North American Electric Reliability Corporation gained binding powers in the United States in 2006, and has advisory powers in the applicable parts of Canada and Mexico. The U.S. government has also designated National Interest Electric Transmission Corridors, where it believes transmission bottlenecks have developed.

Some areas, for example rural communities in Alaska, do not operate on a large grid, relying instead on local diesel generators.

High-voltage direct current lines or variable frequency transformers can be used to connect two alternating current interconnection networks which are not synchronised with each other.

This provides the benefit of interconnection without the need to synchronise an even wider area. For example, compare the wide area synchronous grid map of Europe with the map of HVDC lines.

Redundancy and Defining "Grid"

A town is only said to have achieved grid connection when it is connected to several redundant sources, generally involving long-distance transmission.

This redundancy is limited. Existing national or regional grids simply provide the interconnection of facilities to utilise whatever redundancy is available. The exact stage of development at which the supply structure becomes a *grid* is arbitrary. Similarly, the term *national grid* is something of an anachronism in many parts of the world, as transmission cables now frequently cross national boundaries. The terms *distribution grid* for local connections and *transmission grid* for long-distance transmissions are therefore preferred, but *national grid* is often still used for the overall structure.

Aging Infrastructure

Despite the novel institutional arrangements and network designs of the electrical grid, its power delivery infrastructures suffer aging across the developed world. Four contributing factors to the current state of the electric grid and its consequences include:

1. Aging power equipment – older equipment have higher failure rates, leading to customer interruption rates affecting the economy and society; also, older assets and facilities lead to higher inspection maintenance costs and further repair/restoration costs.
2. Obsolete system layout – older areas require serious additional substation sites and rights-of-way that cannot be obtained in current area and are forced to use existing, insufficient facilities.
3. Outdated engineering – traditional tools for power delivery planning and engineering are ineffective in addressing current problems of aged equipment, obsolete system layouts, and modern deregulated loading levels
4. Old cultural value – planning, engineering, operating of system using concepts and procedures that worked in vertically integrated industry exacerbate the problem under a deregulated industry.

Bibliography

Ashraf, Imtiaz and M.A. Mallick: *Electrical Machines*, I.K. International Pub, Delhi, 2009.

Badiru, A.: *Handbook of Industrial and Systems Engineering*, CRC Press, Delhi, 2005.

Bamberger, Robert L.: *The National Energy Strategy: Summary and Analysis,* Congressional Research Service, Library of Congress, Washington, DC, 1991.

Bard, A.J.; Faulkner, L.R.: *Electrochemical Methods: Fundamentals and Applications*, John Wiley & Sons, New York, 2005.

Barua, Lohit: *Natural Gas Measurement and Control: A Guide for Operators and Engineers*, McGraw-Hill Professional Book Group, London, 1991.

Bauer, M.: *Resistance to New Technology*. New York: Cambridge University Press, 1995.

Bennett, S.: *A History of Control Engineering: 1800-1930*, IEE Press, London, 1979.

Bernard, Finn S.: *The History of Electrical Technology: An Annotated Bibliography*, Garland Publishing, Inc., New York, 1991.

Bird, John: *Electrical and Electronic Principles and Technology,* Newnes, UK, 2007.

Bobrow, Leonard S.: *Fundamentals of Electrical Engineering*, Oxford University Press, Delhi, 2003.

British Electricity International: *Modern Power Station Practice: Incorporating Modern Power System Practice,* Pergamon. 1991.

Brittain, A., and P. Morton: *Engraving on Precious Metals*, N.A.G. Press, London, 1958.

Buckel, W.: *Superconductivity: Fundamentals and Applications,* VCH Publishers, Inc., 1991.

Chakravorty, Jaydeep: *Analysis of Electrical Circuits and Networks*, Laxmi Publications, Delhi, 2009.

Chatfield, C.: *Introduction to Time Series and Forecasting*. Springer-Verlag, New York, 1996:

Christopher Kilian: *Modern Control Technology*, Thompson Delmar Learning, UK, 2005.

Clerc, M.: *New Optimization Techniques in Engineering* Springer,Verlag, NY, 2004.

Davis, H.F.: *Fourier Series and Orthogonal Functions*, Allyn and Bacon, Boston, 1963.

Dhawan, B C : *Textbook of Electrophoresis*, Pearl Books, Delhi, 2008.

Doebelin, E. O.: *Engineering Experimentation: Planning, Execution, Reporting*, McGraw Hill, Boston, 1995.

Dorf, Richard C.: *Electrical Engineering Handbook*. Boca Raton, FL: CRC Press, 1997.

Edminister, Joseph: *Electric Circuits,* McGraw-Hill, NY, 1965.

Eveleigh, E.W.: *Introduction to Control Systems Design*, McGraw-Hill, 1972.

Fabrycky, W.: *Systems Engineering and Analysis*, Prentice-Hall, NY, 2005.

Hammond, Percy: *Electromagnetism for Engineers*, Pergamon, NJ, 1981.

Harriott, P.: *Unit Operations of Chemical Engineering*, McGraw Hill, New York, 2004.

Heisler, H.: *Advanced Engine Technology*. Warrendale, PA: SAE, 1995.

Kuo, B.C., *Automatic Control Systems*, Prentice-Hall, NJ, 1995.

Moorhouse, John C.: *Electric Power: Deregulation and the Public Interest,* Pacific Research Institute for Public Policy, San Francisco, 1986.

Ogata, K.: *Modern Control Engineering*, Prentice-Hall, 2002.

Paramar, Ashwin: *Energy Future : Coping with Critical Conditions*, Dominant, Delhi, 2001.

Randall, Merle: *Thermodynamics and the Free Energy of Chemical Substances.* McGraw-Hill Book Co. Inc., 1923.

Reddy, K. Krishna: *Electrical Measurements*, Scitech Pub, Delhi, 2009.

Ripley, Brian D.: *Pattern Recognition and Neural Networks*, Cambridge, London, 1996.

Singh, R.P. : *Electrical Workshop: A Textbook*, I.K. International, Delhi, 2005.

Tiwari, S.K. : *Fuel Minerals and Other Energy Resources*, Atlantic Pub, Delhi, 2010.

Treager, I.E.: *Aircraft Gas Turbine Engine Technology*. . McGraw-Hill Book Co. Glencoe Div., 1995.

Index

❑❑❑